CLOSE·TO ·NATURE·

CLOSE·TO ·NATURE·

A Naturalist's Diary of a Year in the Bush

JOHN LANDY

VIKING O'NEIL

Viking O'Neil
Penguin Books Australia Ltd
487 Maroondah Highway, PO Box 257
Ringwood, Victoria 3134, Australia
Penguin Books Ltd
Harmondsworth, Middlesex, England
Viking Penguin Inc.
40 West 23rd Street, New York, N.Y. 10010, U.S.A.
Penguin Books Canada Ltd
2801 John Street, Markham, Ontario, Canada L3R 1B4
Penguin Books (N.Z.) Ltd
182–190 Wairau Road, Auckland 10, New Zealand

First published by Currey O'Neil Ross Pty Ltd 1985
This edition published by Penguin Books Australia Ltd 1988
Copyright © John Landy, 1985

All rights reserved. Without limiting the rights under copyright reserved above, no part of this publication may be reproduced, stored in or introduced into a retrieval system, or transmitted, in any form or by any means (electronic, mechanical, photocopying, recording or otherwise) without the prior written permission of both the copyright owner and the above publisher of this book.

Produced by Viking O'Neil
56 Claremont Street, South Yarra, Victoria 3141, Australia
A division of Penguin Books Australia Ltd

Typeset in Australia by ProComp Productions Pty Ltd
Printed and bound in Hong Kong through Bookbuilders Ltd

National Library of Australia
Cataloguing-in-Publication data

Landy, John, 1930–
  Close to nature.

  Bibliography.
  Includes index.
  ISBN 0 670 90090 7.

  1. Natural history – Victoria, Northeastern.
  2. Natural history – New South Wales, Southeastern.
  I. Title.

508.945'5

CONTENTS

Map of the Upper Murray	vi
Author's Note	vii
Acknowledgements	vii
Introduction	viii
The Diary	2
Bibliography	144
Index	147

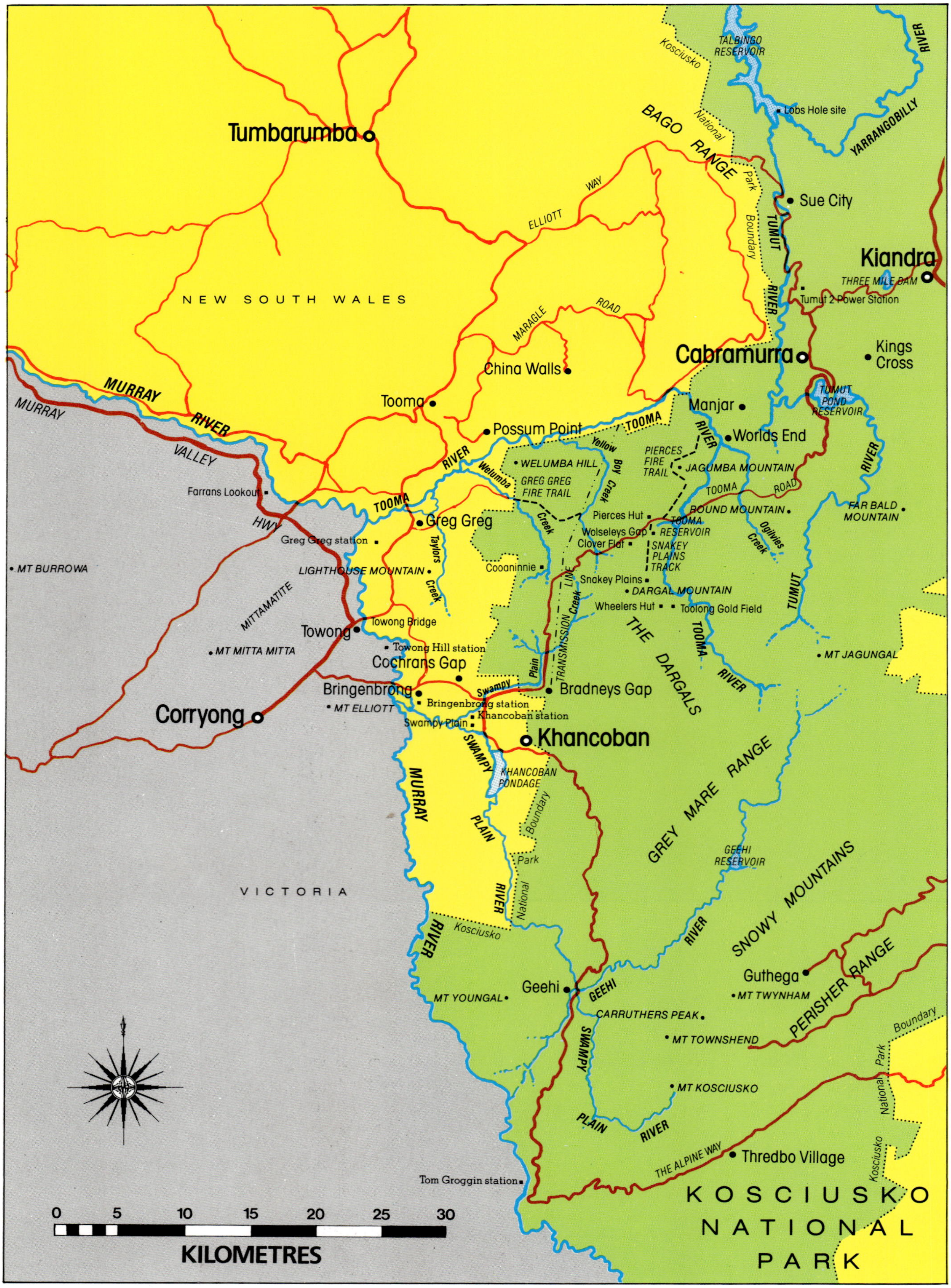
Tumbarumba
NEW SOUTH WALES
VICTORIA
KOSCIUSKO NATIONAL PARK
Kosciusko National Park Boundary
TALBINGO RESERVOIR
Lobs Hole site
YARRANGOBILLY RIVER
BAGO RANGE
Sue City
TUMUT RIVER
Kiandra
THREE MILE DAM
Tumut 2 Power Station
Cabramurra
Kings Cross
ELLIOTT WAY
MARAGLE ROAD
China Walls
Tooma
Possum Point
MURRAY RIVER
MURRAY VALLEY HWY
Farrans Lookout
TOOMA RIVER
Manjar
Worlds End
TUMUT POND RESERVOIR
WELUMBA HILL
GREG GREG FIRE TRAIL
PIERCES FIRE TRAIL
JAGUMBA MOUNTAIN
TOOMA ROAD
ROUND MOUNTAIN
FAR BALD MOUNTAIN
Greg Greg
Greg Greg station
Pierces Hut
Wolseleys Gap
TOOMA RESERVOIR
Clover Flat
SNAKEY PLAINS TRACK
MT BURROWA
LIGHTHOUSE MOUNTAIN
Cooaninnie
Snakey Plains
DARGAL MOUNTAIN
Wheelers Hut
Toolong Gold Field
MITTAMATITE
Towong
Towong Bridge
Towong Hill station
MT MITTA MITTA
Cochrans Gap
TRANSMISSION LINE
THE DARGALS
MT JAGUNGAL
Bringenbrong
Bringenbrong station
Bradneys Gap
MT ELLIOTT
Corryong
Khancoban station
Swampy Plain
Khancoban
KHANCOBAN PONDAGE
SWAMPY PLAIN RIVER
GREY MARE RANGE
GEEHI RESERVOIR
SNOWY MOUNTAINS
Guthega
PERISHER RANGE
MT TWYNHAM
Geehi
MT YOUNGAL
CARRUTHERS PEAK
MT TOWNSHEND
MT KOSCIUSKO
Thredbo Village
THE ALPINE WAY
Tom Groggin station
0 5 10 15 20 25 30
KILOMETRES

AUTHOR'S NOTE

This diary has been compiled from notes written over one year, during visits I made each month to our family property at Greg Greg. Normally I am concerned with the farm itself, and do not stray much beyond its boundaries, but for this exercise I encompassed a range of country from the Murray River to the High Plains near Cabramurra and Kiandra.

Greg Greg, now marked only by an old abandoned schoolhouse, lies close to the junction of the Tooma and Murray rivers in New South Wales, between Corryong and Tumbarumba. The name, which had alternative early interpretations of 'grik grik' and 'coroc coroc', is said to be Aboriginal for frogs, and is appropriate since even today there are many low-lying swampy parts along the Tooma. The district has a rich and varied history, many aspects of which have been admirably covered by the late T. W. Mitchell in his book *Corryong and The 'Man from Snowy River' Country*, by Jean Carmody in *Early Days of the Upper Murray*, and Judy Paton in *Tooma, the Centenary of Education*.

A set of notes such as this cannot hope to be comprehensive in its coverage of the variety of wild life, let alone the geography and agriculture of the area. There are some large 'gaps', notably in relation to birds and animals. The descriptions, the comments and the photographs obviously reflect the way I see things and there is a bias towards creatures small, particularly insects, for which I make no apology.

One difficulty I encountered with this project was that of trying to exactly match the text with appropriate photographs. I was frequently frustrated, missing or spoiling pictures of interesting events which I have written about but consequently have been unable to illustrate. Conversely, a few photographs have been included because they are typical of the area, but rate no mention in the text as they did not provide a 'story' at the time of writing.

For the sake of easier reading I have used the common names for plant and insect species wherever possible. Unfortunately this sometimes may cause confusion where common names vary, as they frequently do with plants, but they can be identified by their scientific names in the Index. Occasionally scientific names were required for rare plant species or to distinguish between similar species. The majority of Australian insects do not have common names so I have used the scientific names where there is no vernacular title. The captions to the photos have scientific names in all but the few cases where insects or fungi could not be positively identified.

JOHN LANDY

ACKNOWLEDGEMENTS

In writing this diary I have received a great deal of help from a variety of sources and wish to acknowledge the following people: The staff of the National Herbarium, particularly Stephen Forbes who identified the majority of the plants discussed and illustrated, and also Rex Filson, Ray Smith and Geoff Garr. Dr Ian Pascoe of the Plant Research Institute at Burnley, who identified some of the fungi I photographed. Dr Ebbe Schmidt Nielson, Curator of Lepidoptera, National Insect Collection, Canberra, who arranged for the identification of insects; Dr Robert Taylor, who provided information on ants; and the various specialist members of staff at the Division of Entomology, CSIRO, Canberra, who made the identifications; and David Kimpton of CSIRO Information Services, who located many references for me. Ken Walker of the National Museum, Melbourne, who identified several species of bees I observed and photographed, and Belinda Gillies, who checked my bird photographs. Various members of the Forests Commission of Victoria who provided a wide variety of information, particularly John Taylor,

OPPOSITE PAGE:
Map showing the Upper Murray and environs.

Dr Geoff Marks, Fred Neumann and Evan Chesterfield. Dr Malcolm Calder and Dr David Ashton of the School of Botany, University of Melbourne, with whom I discussed observations I had made on a number of botanical subjects; and librarian Lois Davey, who so patiently met my many requests for books and papers. Ian Temby of the Fisheries and Wildlife Service, with whom I discussed the problem of native birds on farms, and Dr Phillip Cadwallader of the Snob's Creek Hatchery, who provided me with a wealth of information on the biology of trout and native fish. Tony Powell, Regional Director of the Commonwealth Bureau of Meteorology, who provided early weather records for the Corryong district. Eligio Bruzzese and Madelon Lane of the Keith Turnbull Institute, who answered my queries about weeds and their biological control. Alex Mitchell, former chairman of the Soil Conservation Authority, who personally provided me with much information on the alpine region and gave me access to material in the SCA library; and members of the SCA staff, Warwick Pabst and Dr Harm van Rees. Marilyn Hogben, librarian at ICI Merrindale Research Station, Croydon, who provided data on a variety of agricultural topics. Michael Wood of State Rivers, who furnished me with historical data on the stream flow of the Murray River. Kevin Johnston of the New South Wales Department of Mineral Resources, who tracked down early mining records. Laurie Rogers of Photoclinic, who kept my cameras operating and undertook emergency repair work at very short notice; and Graeme Miller of Latrobe Colourlab, who helped with the production of some of the photographs. In the finalising of the text I am deeply indebted to editors Jane Drury and Helen Duffy, who spent countless hours on this book. I wish to especially acknowledge the written work (full details of which are included in the Bibliography) of P. B. Carne; R. D. Hughes; I. F. B. Common; K. H. L. Key and M. F. Day; I. F. Nolan; P. Atsatt; D. M. Calder; D. H. Ashton; Ian Rowley; A. B. Costin, D. J. Wimbush, D. Kerr and L. W. Gay; E. G. Matthews and R. L. Kitching.

On my numerous visits to the locality I have been greatly assisted by several people. I would especially like to acknowledge the help and hospitality of my brother Richard and his wife, Fiona. I was much encouraged by my discussions with Elyne Mitchell, who has written so well about the Upper Murray and the Snowy Mountains, with Robert Whitehead, who showed me the famous Lighthouse crossing on the Murray which his family tended over the years, and Graeme Whitehead, who found answers to my queries about the origins of local names. Stewart Ross of the Corryong Historical Society was very helpful in providing information and suggesting further sources. Two young farmers whose families are very much part of the history of the locality, Bob Herbert and Terry Pierce, provided me with insights on the seasons, the state of stock and incidence of wild dogs. An old school mate and local property owner, John Montague, took me on a memorable trip over some of his former grazing leases on the High Plains and made old deserted huts and overgrown tracks come to life with his stories of the days of the mountain cattle.

Winter pastoral scene.

INTRODUCTION

The Upper Murray region, straddling the river which forms the state border of New South Wales and Victoria, is one of the most magnificent areas in Australia.

Its lush river flats and rising pasture lands have sustained sheep and cattle since the mid-19th century. Its timbered hills hide waterfalls and creeks, and gullies as they rise to the rocky peaks and tableland of the subalpine region. Higher still, the Snowy Mountains create a white-capped sparkle in the winter and a brooding bulk in summer.

It is a place of confluence in nature—the waterbirds of the Murray and Tooma rivers, the multitude of parrots, cockatoos and common species of woodland and farmland, and the shy forest birds converge on each other's territory. The plant life extends from the stately River Red Gums to the bent and pliant gums above the snow-line, from fern gullies to delicate wild flowers on the subalpine meadows. Emus, kangaroos, possums, wombats, echidnas, wild dogs and feral cats are part of the fauna occupying the various niches of vegetation and terrain and the wide climatic range. Insects, thousands of species perfectly adapted for their special functions in a microcosmic world, are a vital and fascinating sub-culture, helping to weld soil and plant, bird and animal.

It is a place, too, of history and legend. The first settlers worked vast runs centred on the fertile, well-watered flood plains. Miners came to the higher reaches to seek gold and attempt snow sports which presaged the booming ski industry of today. Small towns like Tumbarumba, Kiandra and Corryong came into being to meet the needs of the farmers, miners, labourers, bushmen, artisans and traders who came into the area. The Land Selection Acts of the early 1860s led to the disappearance of the runs and the eventual establishment of the present pattern of intensively managed dairy and grazing properties, and larger stations with a balance of river and creek flats and lightly timbered hill country.

View over Taylors Gully towards the Tooma and Mittamatite.

The practice of grazing cattle on the High Plains in summer extended the range and carrying capacity of the stations, and brought into being a quality of bushcraft and horsemanship which has become part of the Australian ethos. 'Banjo' Paterson's immortal story of 'The Man from Snowy River' had its beginnings here, and horsemanship remains a much admired skill in the region. The development of the huge Snowy Mountains Scheme, which directs the waters of the Snowy River for hydro-electric power and irrigation, helped close the mountains to cattle and provided impetus for the declaration of the Kosciusko National Park.

It is in this area that John Landy took camera and notebook for a close look at nature on a series of visits over more than a year. His base was the family farm and he was involved in its workings through the seasons, as well as ranging widely in the bush on a series of expeditions. From his schooldays, and throughout his athletic, academic and business career, he has been a trained and informed observer of nature, particularly of native insects.

The camera became an essential part of his recording and, as time went by, an instrument for the depiction of evocative images on a wider scale. His knowledge of the area, and his entrancement with it, is evident in his photographs and descriptions. The spectacular terrain is captured in the golden light of summer, the chill shimmer of dawn or the soft mists that lie low in the valleys on still winter mornings. His close focus photography brings us into the tiniest detail of exquisite wild flowers and insects in their many forms of defence and attack, alliance and subterfuge.

OVERLEAF:
The Tooma valley from the lower slopes of the Lighthouse Mountain.

ix

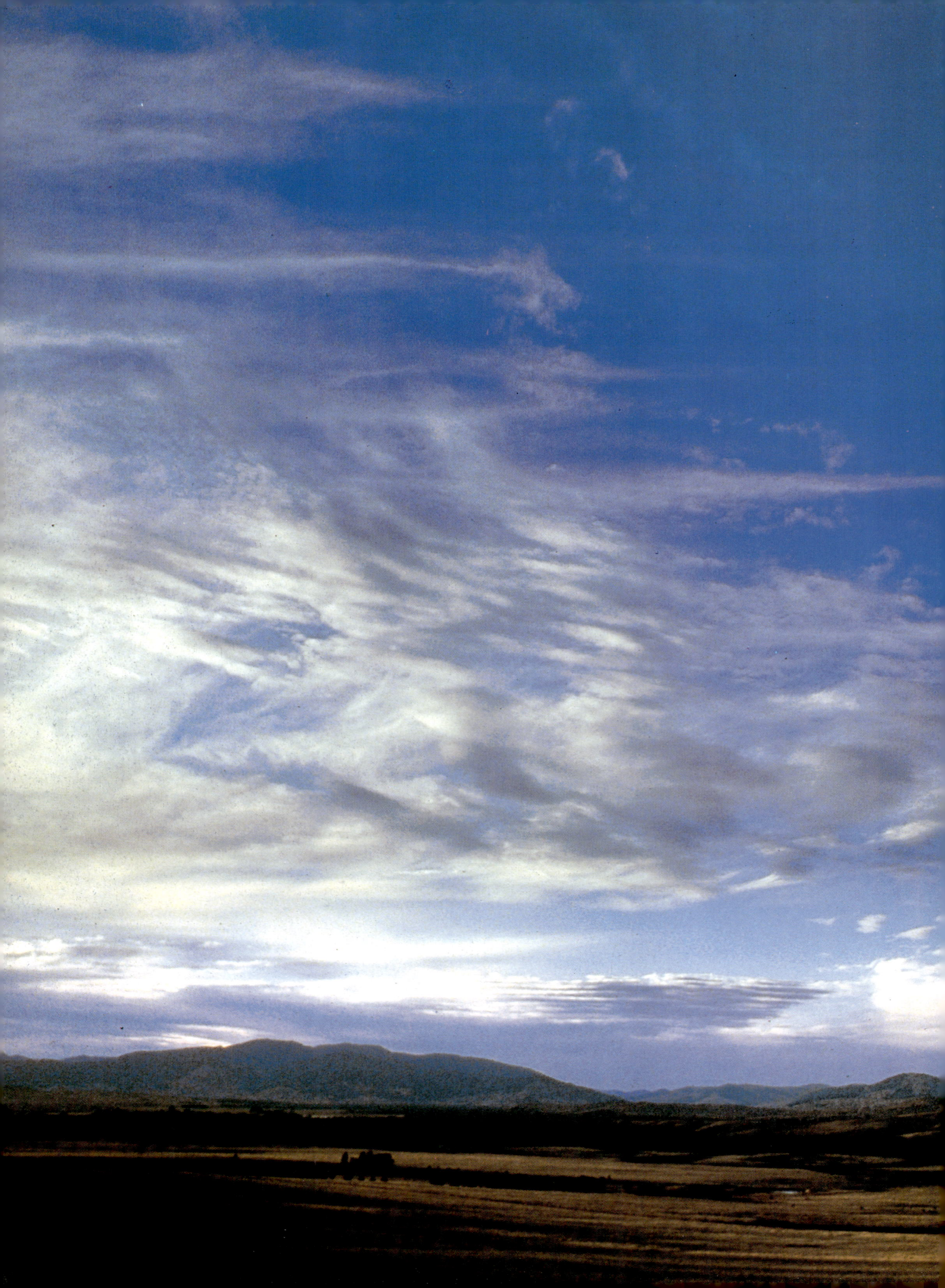

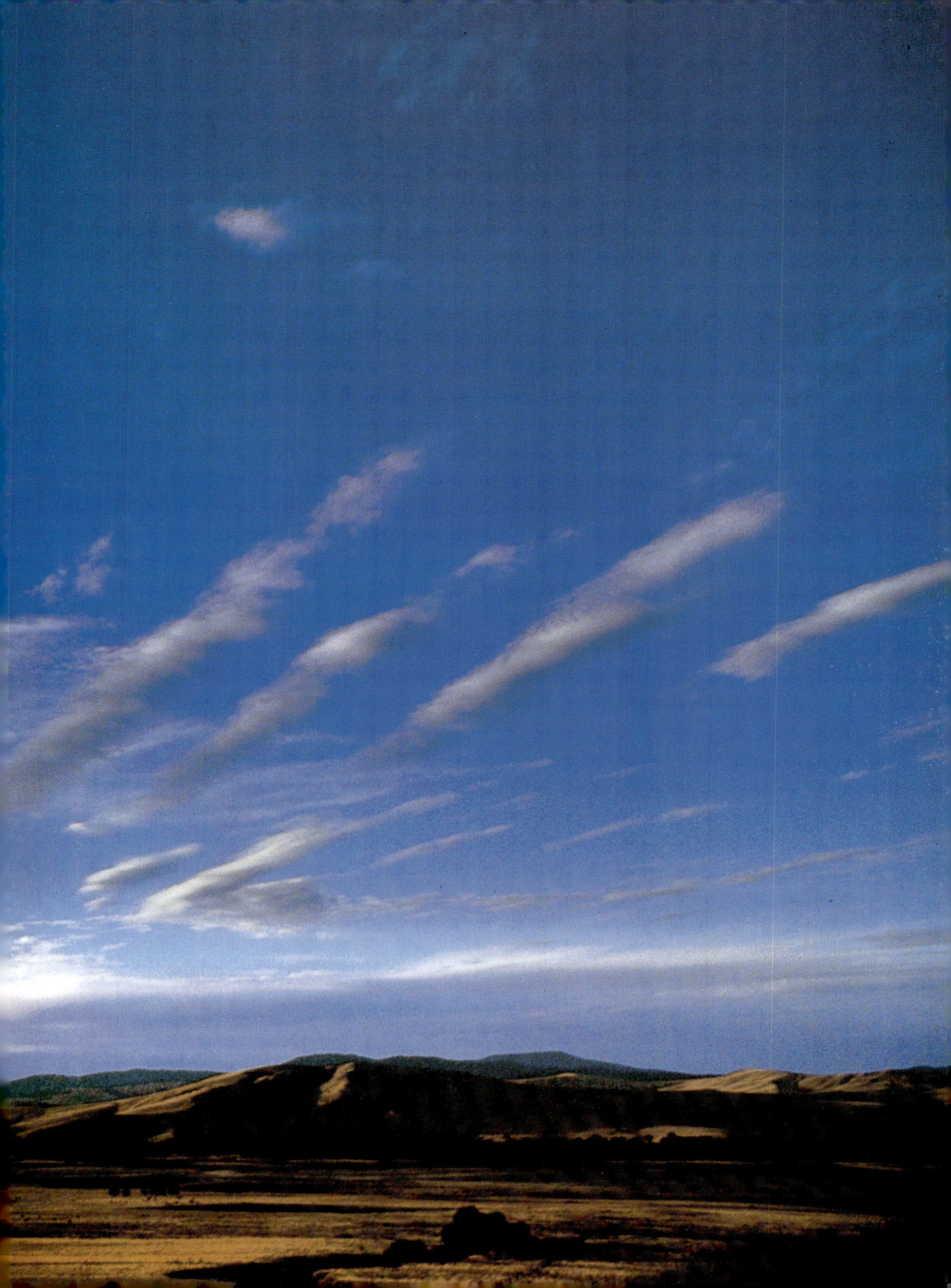

JANUARY 27

I rose at 4 a.m. and set out for the Lighthouse Mountain, a rocky, isolated peak which rises 300 metres sheer above our cottage. This stark and arid survivor of the erosive powers of the Murray and Tooma rivers commands almost an entire sweep of the skyline, blocked only by a sister peak to the south-west.

As I left the cottage the stars were just beginning to fade, giving way to the first rays of dawn light over the Dargals. Early morning here is always a time of great delight to me — with an almost eerie silence punctuated by the bellowing of cattle or the harsh cry of crows, and now with that wonderful smell of dry summer grass.

I expected to take about twenty-five minutes to climb the Lighthouse, but I was slowed down in the semi-dark by the granite boulders and occasional fallen logs. From the summit the ancient valley of the Murray can be seen running up to its source above Tom Groggin. Higher again and to the east is the Kosciusko massif, its precipitous western flank peering out above the forested foothills.

Sitting on the topmost boulder I could pick out the silvery sinuous outline of the Tooma River to the west just before its confluence with the Murray. Light patches of mist blurred the outline of the ancient red gums and the complex system of depressions and billabongs some 350 metres directly below, perhaps much as it was 140 years ago when the valley was first settled.

Before sunrise.

In an old dead gum tree a noisy squabble broke out among a group of roosting Sulphur-crested Cockatoos, always a sure herald of the approaching sunrise.

Sunrise seems to come very quickly on the cleared hills of the Upper Murray. From my vantage point this morning the sun rose over the massive and brooding bulk of Dargal Mountain. The first rays burnished the stems and russet flowers of the Kangaroo Grass, and then expanded to provide an iridescence to new gum shoots and papery thistle heads.

Sun rising over Dargal Mountain, viewed from the Lighthouse.

It was not hard to see how this area has captivated poets, writers and artists. Even the place names of the Snowy Mountains have a melodious and poetic ring. Where else could your eye sweep the horizon and successively reel off names like Youngal, Khancoban, Bringenbrong, Jagungal, Jagumba and Welumba. One understands why Banjo Paterson derived his inspiration for his famous ballad 'The Man From Snowy River' in these hills and valleys.

The area I am to cover in my diary runs roughly east from the Lighthouse Mountain. To the west it includes some of the grassy flats on the Murray River. All in all an area of 800 square kilometres.

It covers a wide variety of terrain. There is country much modified by man since early settlement in the 1840s, including the river and creek flats with their remnant River Red Gums, and the adjoining cleared hills with surviving bushland of Red Stringybark, peppermint and Candlebark. Further east the forests of the foothills give way to the Dargals, with their rocky peaks and dense snow gum regrowth, and to the tableland immediately beyond, sometimes known as the Murray Plateau. This subalpine region forms a mosaic of more rounded peaks, snow gum woodlands and grassy plains, across which the Tooma River meanders north towards Manjar and Black Jack, before plunging at World's End 700 metres to the valley below.

Grass flower (*Stipa sp.*).

JANUARY 28

It has been a wonderful season in the Upper Murray, one of the best the locals can remember. As so often happens, it follows a major drought, of which the dead trees and shrubs are stark reminders. The worst affected are the Silver Wattles. One group I saw today, perhaps fifty trees reaching 10 metres in height, was entirely dead.

Yet the Silver Wattle here is a most resilient tree; it readily shoots again from root suckers and seed, and the conditions which kill the mature plant—fire and drought—provide the stimulus for regeneration. The new shoots, already massing near the base of each dead tree, will reach the height of the original in five years.

Given time, and if no efforts were made to clear it or fertilise the pasture on the cool southern and eastern slopes, the Silver Wattle would reinvade much of the farmland, the new growth being aided by the nitrogen-fixing bacteria on the roots. Clearing of regrowth wattle is difficult as the young trees bend pliantly under the blade of the bulldozer and the roots refuse to budge.

I found one surviving Silver Wattle near a creek, where it had been able to obtain more than its fair share of moisture. However, it would not have many years to live since the trunk was perforated by borer holes and the bark was coming away from one side. It was a veritable hive of insect activity. There were streams of red and purple meat ants running up the trunk and down again, seeking out leafhoppers, scale insects and the butterfly larvae they attend. There were several cleverly camouflaged moth caterpillars, brightly coloured chrysomelid beetles and quaintly shaped weevils feeding on the lacy blue-green foliage.

It is the borers within the trunk—primarily the caterpillars of moths and beetles—which control the destiny of the tree and determine how long it lives. The borers have their problems, too. They are attacked by Yellow-tailed Black Cockatoos which locate the galleries in the trunk or limbs. The cockatoo opens an elongated entrance at the top of the tunnel, works downward and removes the succulent larvae. The success that cockatoos have in detecting their victims suggests some means of sensing the movement or gnawing of the caterpillar, through hearing, or perhaps by the touch of their beaks or tongues. However, they also open tunnels where pupae have emerged or the larvae have been parasitised, indicating that they may detect the presence and outline of the tunnel itself.

Moth caterpillar (unidentified).

Green leafhoppers (*Sextius virescens*) attended by meat ants (*Iridomyrmex purpureus*).

Beetle (Chrysomelidae).

Long-tailed wasp (Megalyridae).

I also saw a long-tailed wasp, a parasite of wood-boring caterpillars. It was flying around the wattle trunk and along the limbs, the white-tipped trailing filaments of its ovipositor almost quivering. These wasps locate their hapless victims and bore through the entrance plug of digested wood particles, and sometimes through the hard timber of the trunk itself, and lay eggs inside their skin. The eggs hatch within the caterpillar, the wasp larvae consume their host and in due course emerge as adult long-tailed wasps.

Some indication of the intense activity within the trunk of decaying wattles was demonstrated some years ago at an international entomological conference in Canberra. No less than 273 insects of twenty-three species were bred from the top 1·5 metres of a Late Black Wattle in a Canberra garden.

Wood-boring longicorn beetle (Cerambycidae) larva.

5

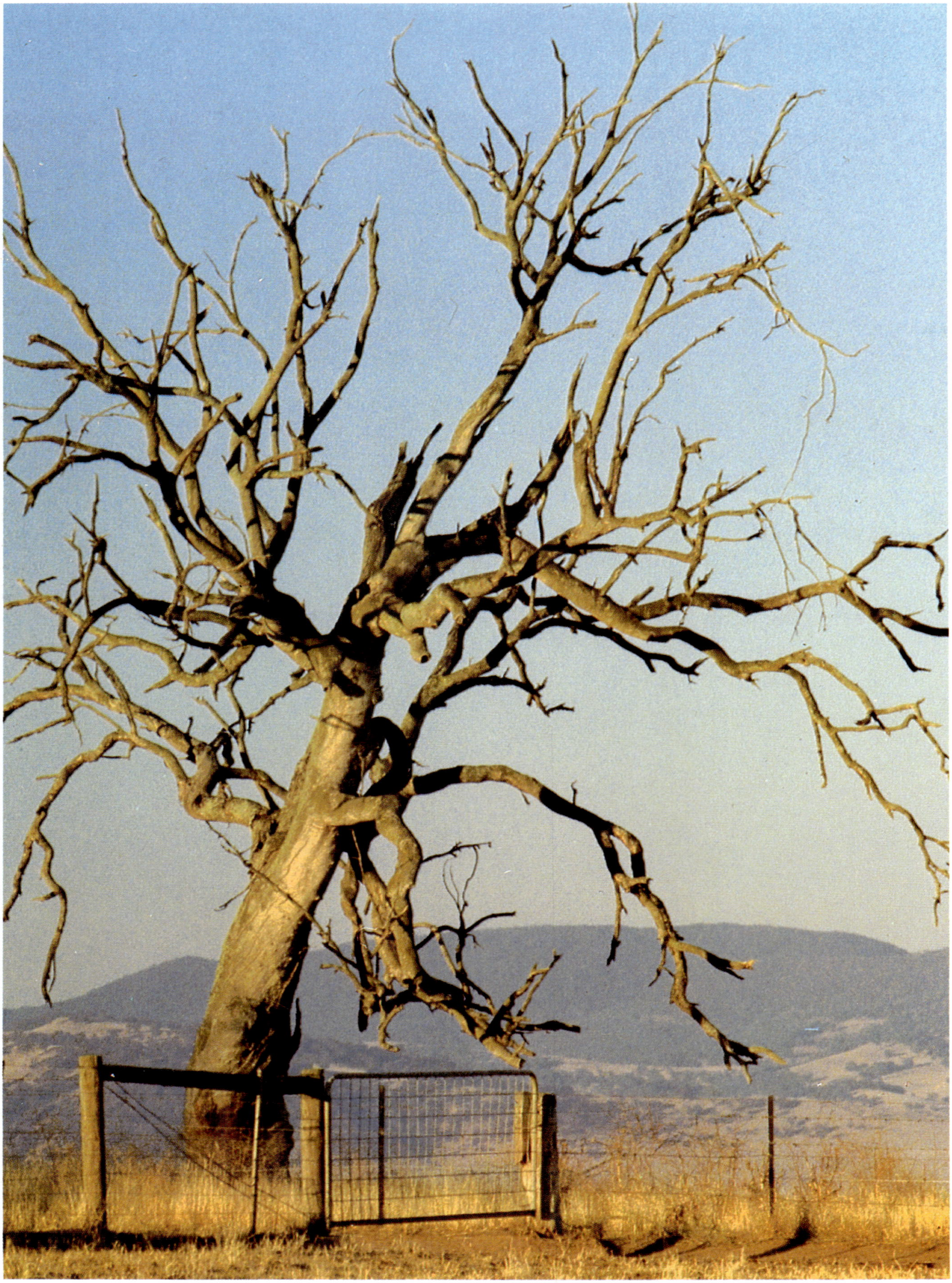

I saw a flock of Emus in a paddock adjoining the Kosciusko National Park. We often see them here as they like to feed on the verge of the forest, where there is a greater variety of food. They were not at all tame and soon ran swiftly away. Shortly afterwards I saw an adult with half a dozen young birds, easily identified because of their smaller size and attractive striped markings. The adult was probably a male, which incubates the eggs and tends the young.

Emus, like kangaroos, have probably been encouraged by the development of improved pastures, which provide a source of nutritious grass and seed heads superior to that obtainable in the open forest, and also more insects, such as grasshoppers and caterpillars. The farmland includes patches of blackberry and Sweet Briar and the fruit of these two weeds forms part of the Emus' diet. Although the Emus are not a serious problem—unless they get caught in the fences—they may have contributed to the increase in blackberries. I have noticed the appearance of new blackberry bushes away from the gullies and it seems possible that they have been spread in Emu droppings.

I stopped briefly at the side of the clear mountain stream that flows through the paddock. This provides water for the homestead and stock. At a flatter section, under the shade of some willows planted many years ago, there is an accumulation of coarse gravel and sand. On this, just next to the running water, a Meadow Argus butterfly was preening its wings and seemed to be drinking from the wet sand. In the tropics a frequent and brilliant sight are the scores of brightly coloured butterflies grouped around puddles after rain. In Australia butterflies are often seen at flowers seeking nectar, but it is not so common to observe them drinking at the side of puddles or streams.

This butterfly was easily recognised as a member of the Nymphalidae, a family which has lost the locomotive function of the forelegs which are held tucked up on the underside of the thorax. I disturbed the butterfly but it returned to the same spot and tentatively coiled and uncoiled its proboscis like a watch spring.

Highly sensitive receptors are located on the tip of the proboscis and, strangely, on the tarsi or feet of the middle pair of legs. So perceptive are these organs that some species can recognise sugar in solution at $\frac{1}{250}$ the level of human detection.

Most insects have been scarcer this year—a consequence of the drought. One exception is the Black Cicada and the noise from these insects at times is overwhelming. Today, in one small section of the forest, there were literally hundreds of them. On one tree trunk alone I counted half a dozen, some of them quite near the nymphal shells from which they had emerged earlier in the day. Later they will climb the tree and begin sucking the sap with their powerful beaks.

This jet black species has bright red eyes and is about 5 centimetres long. Black is not a usual colour for a diurnal insect and it is possible that with the many fire-blackened trunks it confers some measure of concealment, though I have seen the cicadas resting on unburnt bark.

One reason the Black Cicada has not been affected by the drought is its longer life cycle than most other types of insects. It spends several years in the immature stages living in a moist sealed chamber beneath the ground.

It is the male in the cicada family that produces the sound and which prompted the ancient quotation, 'happy the cicada lives for they alone have silent wives'. The sound is produced by the rapid vibration of the tymbals or drums on the first segment of the abdomen. The organs of hearing are not so obvious. Their identity was one of the puzzles of entomology, not solved until the early 1920s, when it was shown that the hearing organ was a mirror-like object located beneath the sound-producing organ. Strangely, the ear of the female cicada, to whom the love-song is directed, is smaller than that of the male. It is known that the males compete against each other with their song, and in fact reply to each other's calls.

I came across a small group of kangaroos. They were Eastern Greys, a large species with the male attaining a height of 2 metres. The Eastern Grey is the common kangaroo of south-eastern Australia—the coastal and highland areas in particular. We see them on the property during summer, but they are found in big numbers during the winter months when feed is short. At the moment there is such an abundance of feed everywhere that they can find plenty to eat in the forest.

Emu.

OPPOSITE PAGE:
Midsummer morning heat haze, across Tooma valley towards Mittamatite.

Meadow Argus butterfly (*Junonia villida calybe*).

Kangaroos, like wild dogs, cause friction between the farmer and the conservationist. The problem of the kangaroo is often not so much the pasture they eat, but rather the damage they do to fences. Contrary to popular belief, kangaroos do not readily jump fences, but make holes to go through, even in well-laid wire netting. It is virtually impossible to prevent kangaroos from getting through wire netting once they have established their pathways. In country where wild dogs are a problem, they can make use of these holes to prey on sheep and lambs.

Late afternoon clouds, western flank of Lighthouse Mountain.

JANUARY 29

Again today was cool. Usually we expect many days of 30° or more during January and February but this weather is more like autumn or spring and, combined with the excellent rains, has lent a pattern of greenness to the January scene quite rare in my experience. The great wide valleys of the Murray and the Tooma are usually characterised in summer by harsh, dry, shimmering pastures which reflect the radiant heat from the sun. It always surprises me how often Corryong appears with the highest temperature recording for Victoria and yet (in a straight line) it is situated not much more than 40 kilometres from the summit of Mount Kosciusko. The mild weather this year has been welcomed by most farmers, but there are some who find it worrying. Those trying to harvest grass seed are being held up until the crop ripens, while others are concerned about flies striking their sheep.

A few struck sheep were being cleaned up at the stockyards this morning. In Australia probably no other animal health problem has been given so much attention as sheep blowfly and it is still estimated to cost the industry $150 million a year. But considerable progress has been made and it is not as serious as it was in the 1950s when vast numbers of sheep died during outbreaks.

The primary infection is most often caused by a small metallic green fly introduced from South Africa late last century. Other species of flies are also involved, particularly in secondary infection. The young maggots of the primary fly are only able to take advantage of sheep skin already damaged by some other cause, such as bacterial fleece rot or scalding by urine or faeces.

The sheep blowfly has shown exceptional adaptability and has become resistant to insecticides, though the strategic use of chemicals still remains a most important element of the control programme on a farm.

Over the years the approach has broadened to place increasing emphasis on keeping sheep clean. The mules operation to remove folds of skin from around the crutch has proved most effective. It is a gory operation carried out on young sheep but it does them no harm. Crutching each year around the breech area helps, as does regular drenching against stomach worms to prevent scouring. Nevertheless, even well-cared-for sheep get flyblown. If there is a humid period during the summer wool becomes moist and bacterial infection can occur and attract flies.

Research has taken several new directions in recent years. The most intriguing and, theoretically, the one which holds the possibility of eradication, is the manipulation of the genetic makeup of the fly—for instance, to render it less fertile, or more susceptible to insecticides. Though with so wide a distribution and the sheer numbers of flies in Australia it seems unlikely that the pest could be entirely eradicated. Other promising lines of investigation include the development of vaccines to confer immunity on sheep against the blowfly or bacterial fleece rot, and the production of viruses and bacteria. Although the flies have outwitted the scientist to date, much more sophisticated and effective weapons seem likely to be launched against them in the near future.

I spent some time this morning looking at semi-cleared areas on the margins of the property. No accurate records exist as to when this property was first cleared, but it formed a part of the Greg Greg run and some clearing may have been carried out in the 1850s. I wanted to see what introduced plants have become established, particularly where forest joins grassland. Two weeds stood out because of their colour, St John's Wort and Paterson's Curse.

St John's Wort is an attractive yellow-flowering plant with minute black glands bordering the petals. The flower contains a red pigment from which the name St John may have been derived as an allusion to his blood. Late last century it was brought to north-east Victoria for medicinal purposes, by a German woman who practised as a midwife in Bright. The plant is said to have escaped from her garden to become a serious weed in the mountain districts of Victoria and southern New South Wales and was first observed near here at Tumbarumba in 1899. Its movement was initially encouraged by the gold-rushes, with seed carried in the chaff for the miners' horses. It is very prolific and a single plant may produce up to 30 000 seeds which can remain dormant for several years.

St John's Wort tends to find a niche in forest country which has been partially cleared, heavily grazed, affected by rabbits, or where the ground has been disturbed by mechanical equipment. You don't see it in highly improved pastures where competition from the more vigorous introduced plants contains or eliminates it. Biological suppression of this weed is possible and eight species of insects were introduced for this purpose many years ago. Of these, one in particular, a beetle and a natural predator of St John's Wort in Europe, continues to exert some control. In more remote and isolated parts it has not been successful because plants germinate again from dormant seed and because the beetles have limited mobility. The plant is toxic to stock, causing photosensitivity or wort dermatitis.

Paterson's Curse has a brilliant mauve flower and often clothes whole hillsides in a blaze of colour. It was first recorded on a property near Albury in 1888 but is said to have been brought to the district as a garden plant by a Mrs Paterson some twenty years earlier. It invades both improved and unimproved pasture and is difficult to eliminate once established. The plant has some low nutritive value for stock and in the more arid regions of South Australia paradoxically goes by the name of Salvation Jane.

Recently Paterson's Curse has been the centre of controversy. Graziers would

Sheep blowfly (*Lucilia cuprina*).

St John's Wort (*Hypericum perforatum*).

One of the species of beetles (*Chrysolina sp.*) introduced for the biological control of St John's Wort.

Paterson's Curse (*Echium lycopsis*).

Paterson's Curse invading newly sown pasture.

like to see it eradicated or at least suppressed and the CSIRO have several suitable insects for release as biological control agents, including beetles, moths and plant bugs, all predators of the plant in its natural habitat in the northern hemisphere. Bee-keepers, however, see Paterson's Curse as an important source of honey.

Unfortunately last season's [1982–83] drought provided an opportunity for the spread of Paterson's Curse into bare patches on the pasture land. It is now much more common and will be difficult to eradicate. It can be controlled by the application of suitable herbicides but the timing has to be right and complete elimination of the plant is made difficult by later germination of dormant seeds.

The designation of a plant as a weed is relative. A weed is really a plant growing out of place and only wins the name when it becomes a nuisance. In biological terms it is a very successful plant — able to adapt to the biological niche that has been opened to it. Few weeds are able to invade Australian bushland unless it has been modified, and they succeed particularly where the change has been gross, either by heavy grazing or by removal of the original vegetation. Some weeds can compete successfully against pasture species that are introduced later. The blackberry is in this category.

Since its introduction in the 1850s the blackberry has spread throughout southern Australia, occupying farmland and uncleared crown land. Nine species of this member of the rose family are present in Victoria and New South Wales.

It is sometimes said that Baron Ferdinand von Mueller, Australia's premier botanist, introduced the blackberry. Rather it seems that the Baron promoted its use in articles he wrote stressing its value in the control of creek bank erosion and as a fruit. Certainly its potential as a weed was not recognised. It was also an age in which societies were formed for the specific purpose of introducing 'valuable' plants and animals from the northern hemisphere.

Today the blackberry occupies 3 million hectares in Victoria alone and probably costs Australia up to $40 million annually. It can be eradicated by spraying, although it is usually necessary to repeat this each year for four years or more. But the real difficulty in controlling blackberry is that so much of its present habitat is virtually inaccessible. It has invaded remote farmland and spread into national parks, and gullies and clearings around old abandoned mining settlements now largely returned to bush. The main thrust of research is now biological control.

I saw two Wedge-tailed Eagles lazily circling the Lighthouse Mountain later this morning. They probably would be looking for rabbits. We have been conducting a rabbit control programme for many years with myxomatosis and 1080 poison (sodium fluoroacetate). By utilising the European Rabbit Flea, a more successful and consistent vector than the mosquito, and with a new strain of the virus, myxomatosis has been more potent and reliable than it was some years ago.

It was significant this morning that there were crows flying with the eagles, as they are always on the scene very quickly when any dead or dying rabbits are about.

I am curious about crows and the way they communicate with each other. When poisoning rabbits years ago I was amazed that, no matter how early I got up in the morning, I would always find gutted bodies of rabbits. Before laying the poison I would not see one pair of crows in a 'day's march', but the next morning there would be literally hundreds. How do they summon each other to the site of the feast from what must be very considerable distances? This is all the more remarkable since it seems to be done before first light. How do they locate the dead rabbits in the dark or pre-dawn light in the first place? The clue may be in the roving groups of non-breeding birds with many pairs of keen eyes, and their raucous cries which can attract colleagues from far afield.

The other thing which surprises me about crows is their relative immunity to the very potent 1080. It is known to be less toxic to birds but crows seem to eat the intestines, where the concentration of poison should be greatest.

Contrary to popular belief crows do not attack healthy lambs or sheep. They only attack—and cruelly—when the sheep are ill, injured or dead. The diet of the crow, particularly in summer, would seem to be made up largely of insects, and at other times of the year plant material forms a significant part of their food source.

After lunch I forsook the farmland to look at some of the country above 1200 metres. It is difficult and time consuming to reach these parts on foot, so I drove up using the sealed road behind the property that runs to Kiandra and on to Cooma. The Snowy Mountains Authority has cut what should prove a very useful swathe through the country for the transmission line, which follows the road for

Late summer afternoon sky.

some distance and provides easy access to a number of distinct vegetation types. On the way up from the Murray River to the High Plains I passed through examples of most of the significant vegetation units which occur here.

The Murray River flats still retain a few remnants of the original savannah woodland, dominated by River Red Gum, and the adjoining gentle slopes have only occasional examples of Blakely's Red Gum, which with Apple Box once were distinguishing features of these mainly red duplex soils with their clay subsoil. The neighbouring steep hills carry typical dry sclerophyll forest of Red Stringybark and in more sheltered aspects Broad-leaved Peppermint, Candlebark and Narrow-leaved Peppermint.

At higher elevations the road passes through tall woodland of Mountain Swamp Gum and then quite suddenly into a patch of wet sclerophyll forest dominated by Narrow-leaved Peppermint, with a lush understorey of Blanket-leaf, pomaderris, Blackwood and Silver Wattle.

River Red Gum (*Eucalyptus camaldulensis*) on the Murray River near Bringenbrong.

Red Stringybark (*Eucalyptus macrorhyncha*) with an understorey of Kangaroo Grass (*Themeda australis*).

Mountain Gum (*Eucalyptus dalrympleana*).

The final stages of the climb pass at first briefly through a more stately and orderly forest of Alpine Ash, clean white boles emerging from shaggy butts, and a uniform understorey of Hop Bitter-pea, and then to the snow gum and Mountain Gum woodland on the crest of the range.

I stopped at Ogilvies Creek, a tributary of the Tooma River and just beyond the Tooma Dam. What a contrast with the farm, only 15 kilometres away. The climate here was quite mild, even on this hot summer day. Ogilvies Creek flows through a

Orange Everlasting (*Helichrysum acuminatum*).

Subalpine flowers: White Chamomile Sunray (*Helipterum anthemoides*) and Billy Buttons (*Craspedia glauca* sensu lat).

Pale Everlasting (*Helichrysum rutidolepis*).

subalpine snowgrass meadow, the blue-green snowgrass providing a background to the dense groupings of yellow and white everlastings. In places the plain is broken by small swamps containing sedges, and on less swampy areas masses of Billy Buttons, each a perfect yellow sphere on a tall, upright green stem. Along the creek there is overhanging growth of prickly Swamp Heath, in places almost hiding the rushing water from view. In sharp contrast the ridges surrounding the plain are covered with snow gums and the verges with a variety of plants, including bright blue-flowering Diggers Speedwell, magenta trigger-plants, everlastings and orange Alpine Oxylobium.

January 30

Last night it rained very heavily — a typical summer thunderstorm. They are common here, providing little useful rain, frequently damaging tracks, causing flash floods and eroding steep land. There was still plenty of cloud this morning and the ground was very wet, and as it dried out the atmosphere became steamy.

There were dozens of Willy Wagtails around the cottage, feeding on the grubs, grasshoppers and beetles that the rain seemed to have stirred up in the pasture. There were plenty of other birds, too — the ever-present Sulphur-crested Cockatoo, a flock of Galahs, and Red-rumped Parrots.

Sunset ahead of approaching summer thunderstorm.

All these birds have benefited from the clearing of the original open forest. The pastures and crops now provide a source of food not previously available, and they have consequently bred up. The Galahs and Sulphur-crested Cockatoos are classed as pests by many farmers.

The male Red-rumped Parrot is brightly hued and the female more subtly coloured, but both are surprisingly well camouflaged when feeding on the ground. The flock around the cottage is very wary at present but you can approach while the birds are eating.

I also noticed some Rainbow-birds for the first time this year. They are striking birds, with a background colour of iridescent green, embroidered with red, blue and black markings around the head and the neck and some flashes of blue on the tail. I saw a group of six on one side of a gully, next to the cottage, and discovered that they were near a nesting site, marked by a tunnel half way up the gully wall, well located to afford protection from snakes, lizards or any animal predators. There will be a nesting chamber within 1 metre of the entrance.

I was leaving for Melbourne this morning, but had time to visit the breeding site of the Ictinus Blue butterfly, which has frequented the same Silver Wattles on Taylors Creek for sixteen years, to my knowledge. This butterfly, like many species of blues (Lycaenidae), has an association with the meat ant—the familiar mound-building ant with red and purple colouration. Meat ants do not sting, unlike the larger bull ants, but are capable of giving quite a painful bite. Their name comes from their alleged habit of stripping carrion. They are also important associates of other insects, such as this blue butterfly, and it is a bizarre sight to see the delicate butterflies fluttering around a seething mass of ants and their own caterpillars. The relationship has often been accepted as a mutually advantageous or symbiotic one, the pugnacious ants giving the caterpillars protection in return for the reward of honeydew secreted from a dorsal organ on a rear abdominal segment of the caterpillar. The defence, from a human being's point of view, is effective and immediate. Should you try to remove a caterpillar or pupa from a branch of the wattle food plant you will be instantly bitten by a host of meat ants which, if given any chance, will extend their interest to all parts of your body! The meat ants' services are also extended to other insects and I found immature stages of Green Treehoppers and scale insects being tended by them.

However, while this idea is appealing it seems rather too simple. Not all species of lycaenid larvae attended by ants produce honeydew, nor are those attended free of parasitism from wasps or flies, and other explanations of the association have been sought.

Another quite different hypothesis suggests that the caterpillar (and the pupa, too), through secretions from glands situated on the abdominal surface, is able both to attract the ants and curb their natural aggression. Although much remains to be discovered about the function of these glands, a likely explanation is that the secretion is a pheremone resembling those produced by the ants' own larvae.

The behaviour of the lycaenid larvae also seems designed to limit the aggression of the ants, because unlike other caterpillars which react violently to the presence of ants, they remain quite passive.

Furthermore, lycaenid caterpillars are protected by an extraordinarily thick skin, or cuticle, some thirty times thicker than that of other types of caterpillars of comparable size, and the jaws of the attending ants, unlike some other insects and spiders, are so constructed that they are unable to pierce it.

Ictinus Blue butterfly (*Jalmenus ictinus*).

Pupa of Ictinus Blue butterfly attended by meat ants (*Iridomyrmex purpureus*).

FEBRUARY 23

In the three weeks since I was last at Ogilvies Creek the flowers have changed; some of the alpine daisies have waned and others are taking their place. From now on the flower display will begin to fade. The season is short in these subalpine regions, beginning in November, reaching its peak about mid-January and ending in March.

In the higher alpine areas of the Kosciusko National Park one species, the Alpine Marsh Marigold, actually begins flowering beneath the melting snow-drifts which may last into December in some years.

The patches of flowers among the snowgrass, mainly Orange Everlastings, Pale Everlastings, Scaly Buttons and Billy Buttons, were attracting a horde of insects. Although they were in great number, there were only a few species—some butterflies, a few bees and carpet moths (*Euphyia*).

The butterfly, with its long proboscis, is ideally suited to gather nectar from these subalpine daisies and there were masses of small silver-stippled browns (*Oreixenica*) and fast flying skippers. Even more prominent were the carpet moths, so named because of their patterned forewings which contrast with the plain hindwings, often yellow or orange in colour.

What struck me today was the great similarity of the moths and butterflies in their flight and their modes of settling on the flowers and extracting nectar. Although not closely related they have developed along similar evolutionary pathways as pollinators of the same group of flowers.

Many of the flourishing young snow gums were being attacked by insects. The thick, broad, bluish-green leaves seemed a highly favoured food of caterpillars, beetles and plant bugs. As usual the most obvious of these were sawfly larvae. They were very active even before I came close, seeming to sense my presence and making convulsive jerky movements. They are often known as spitfires because when alarmed they exude a fluid strongly smelling of eucalyptus. The tightly knit mass of larvae also rears up suddenly, expanding its size and presumably scaring away curious birds and animals.

Sawflies are a primitive form of wasp and in the larval stages rely very much on surviving by living as a group. Individual sawfly larvae, separated from the cluster, do not develop, and languish in the absence of their fellow creatures.

Introduced honeybee (*Apis mellifera*) on Orange Everlasting (*Helichrysum acuminatum*).

Satin Everlasting (*Helichrysum leocopsideum*).

OPPOSITE PAGE:
Billy Buttons (*Craspedia glauca sensu lat*).

Macleay's Swallowtail (*Graphium macleayanum macleayanum*).

Correae Brown butterfly (*Oreixenica correae*) on Billy Buttons.

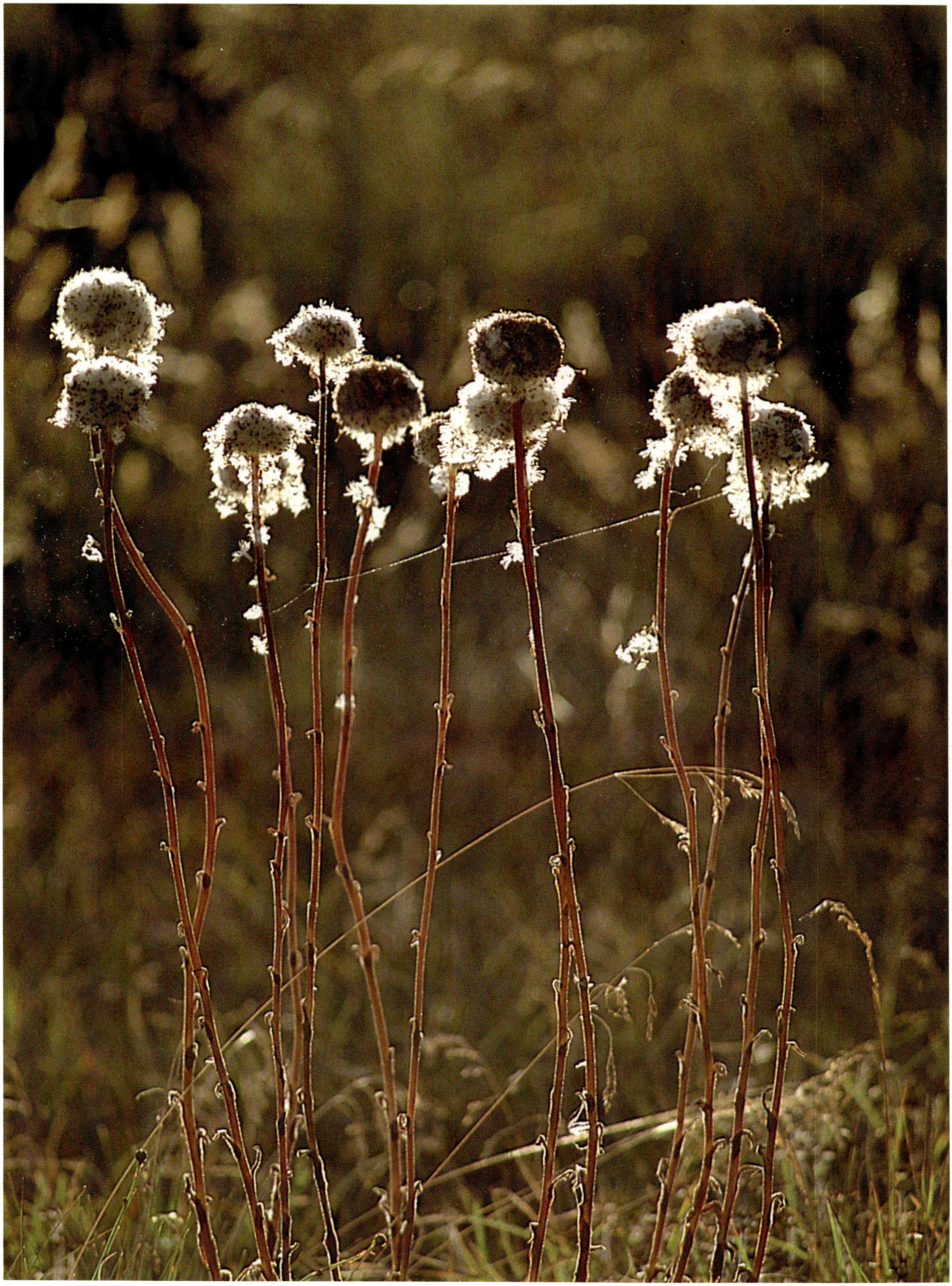

Sawfly (*Perga affinis*).

Sawfly larvae (*Perga sp.*) on snow gum (*Eucalyptus pauciflora*).

Ruby Bracket Fungus (*Tyromyces pulcherrimus*) on the trunk of a snow gum (*Eucalyptus pauciflora*).

Experimental work on one species (*Perga affinis*) has shown that the survival rate of larvae from emergence to pupation is much influenced by the size of the group. The female lays the eggs in a pod running parallel to the mid-rib of the leaf, which is cut between the leaf surfaces, the eggs lying at right angles to the mid-rib. A long preliminary period is spent abrading the leaf surfaces, presumably making it easier for the larvae to escape. The greater the number of larvae emerging from the eggs, the more likely there will be leaders to establish the escape route. They do not escape separately but appear to rely on one or two vigorous individuals to breach the leaf epidermis and lead them out. (Similarly, it has been shown that certain individuals tend to lead the nightly feeding forays.)

Eventually the larvae leave the tree and seek out patches of bare ground in which to bury themselves and pupate. Again the number of larvae which survive to pupate is much greater in large groups than in small ones, since some of the larvae on the outside of the cluster die from desiccation as they excavate the hole.

It is not surprising, then, that contact and the means of maintaining it are at a high premium. Sawfly larvae are constantly seeking each other by tapping the larvae immediately behind with the hardened tip of their abdomen. When an individual becomes separated from the group it raps on the leaves or branches. This quite audible sound elicits an immediate response from the others and eventually the isolated individual finds its way back to its fellows.

Prior to pupation several separate groups may coalesce into huge masses numbering hundreds of larvae, and sometimes are joined by different species of sawfly larvae.

One painful discovery I made today was the presence of some vicious march flies. These green-eyed pests, which have no hesitation in plunging their proboscises through your skin and will bite through your socks and pants, lurk in the shade of snow gums. I cannot see why they do not frequent the open spaces, unless it is that they have become adapted to the habits of mammals of various species which shelter in and among the trees during daytime. I imagine they must have had a 'field day' when cattle grazed here twenty years ago. Perhaps these days they take it out on the occasional kangaroo or wallaby, but it must be a long time between drinks!

I saw a spectacular fungus at the base of an old snow gum. It was a species of bracket fungus—the Ruby Bracket (*Tyromyces pulcherrimus*)—and a brilliant, almost iridescent crimson. It belongs to the polypores, fungi mostly without stems and with pores rather than gills, unlike the common mushroom and toadstool.

The part of the fungus I photographed was the fruiting body. Although the snow gum looked healthy enough a portion had been invaded by thread-like cells or hyphae of the fungus, which in all likelihood will kill the centre of the tree. In time the tree will die. This is a fairly rare fungus and in my experience is only found in a cool climate—synonymous with high places at this latitude.

Some of the polypores belonging to the genus *Fomes* have a perennial fruiting body. They assume huge proportions on the trunks of eucalypts and even develop growth 'rings' much the same as the tree itself. When ignited they burn for many hours and purportedly were used by the Tasmanian Aborigines to transport fire from one camp site to another.

I returned to the farm later this afternoon to learn that a wild dog had been shot. It had been killing sheep for some time but until now has proved elusive. It was definitely not a Dingo, although its appearance suggested some portion of Dingo in its pedigree. The Dingo and the wild dog are members of the same species and so readily interbreed. In fact 75 per cent of all so called Dingoes caught in the south-eastern highlands are believed to be domestic dog crosses. I suspect the proportion of crosses found on the interface with farmland would be much higher again.

One of the few remaining sections of the old dog fence that once separated farmland from the National Park.

Wild dog near Clover Flat.

The wild dog problem is a serious one here, and the National Park provides them with a huge refuge. For years there may be little trouble, then the dogs will go on a killing spree. Some time ago we lost several hundred sheep in the space of about eighteen months, and during this period nine dogs were shot on the property. No sound of the dogs would be heard during the evening and yet they would come right up to the house at night and attack.

They can be stealthy, cunning killers, often taking no more than the kidneys from the bodies of their victims, and sometimes nothing. It would appear that once a dog goes on a rampage of killing it can set up a blood-lust among a group of dogs.

Local farmers are worried about dogs at the moment because of good seasonal conditions. This accelerates the breeding rate and already there are far more of them in evidence than is usual in the National Park and on its fringes. Crossbred dogs are considerably more fertile than Dingoes, having two breeding cycles within one year.

The whole question of wild dog control is a vexed one. In other districts farmers have installed specially designed electric fences with some success. In the absence of a dog-proof boundary fence around the park, for total control the whole boundary of a property must be fenced in this manner. The boundary here between park and farmland was fenced with a 2-metre-high dog fence by the government during the Depression of the 1930s, but this has fallen into disrepair and the cost of replacement today would be high.

FEBRUARY 24

Early this morning, in the mountains, I found a patch of dead snowgrass a few square metres in extent, in stark contrast to the healthy blue-green plants around it. Closer inspection revealed that it had been attacked by underground grass grubs, which are the caterpillars of the swift moth, a native species. The caterpillars live in tunnels beneath the grass tussocks and emerge at night to feed on the leaves. Like other native insects they are usually kept under tight control by predators, parasites and a variety of environmental factors. Occasionally, however, they break out of these restraints, breed up rapidly, and cause damage to quite large areas of snowgrass. The consequences can be serious because of the very slow recovery of alpine and subalpine vegetation.

I also noticed some native bees (*Leioproctus*) burrowing busily into moist soil along the roadside to build their nests. They appeared to be wearing bright yellow socks on their rear legs—the masses of pollen grains they had accumulated from flowers! Bees collect the pollen in special organs called scopa, usually, but not always, on the hind legs.

Like all bees they were very industrious, but after a while each bee would seemingly give up, then start afresh. They were accompanied by some wasps (*Labium*) which appeared to be doing much the same thing but were less persistent. The wasps would be parasites of the bees' larvae, but in the strange 'programmed' manner of insects the bees took no notice of them as they toiled.

Despite the well-known and sophisticated social organisation of introduced bees, most Australian species are in fact solitary and build their own nests without help from separate castes. Bees are really specialised wasps, closely related to sphecoids, the spider and insect hunters. However, bees use pollen rather than spiders and insects to provide protein for their larvae and are dependent on the nectar of flowers for their supply of carbohydrate. Australia, with an estimated 3000 of the world's 20 000 species of bees, many of them still undescribed, is unique among the continents in that most of the species here are dependent on a single family of plants, the Myrtaceae, primarily eucalypts and tea-trees.

Shortly after my encounter with the burrowing bees and the wasps, I came across a small gully containing some patches of sphagnum moss. It is an attractive plant, its closely interlocking leaves forming a dense pale yellow-green mat. Today I found a number of small plants growing out of the sphagnum, including the white-flowered Mountain Gentian.

Over a long period of time sphagnum and its associated herbs and woody plants create deposits of peat which may be more than a metre in depth, and dense enough to significantly raise the water table above that of the immediately surrounding countryside. Sphagnum has a great capacity to hold water, claimed to be as much as twenty-four times its own weight. Today I squeezed half a cup of water from a small section next to some rocks.

Sphagnum and the related ground-water communities, situated across drainage lines and at the source of springs and occupying less than 10 per cent of the high

Native bee (*Leioproctus sp.*) showing pollen-bearing scopa on hind legs.

Wasp (*Labium sp.*), a parasite of the bee (*Leioproctus sp.*).

country, have a significance much greater than their extent would suggest. They perform an important regulatory function since nearly all streams in the high country flow through them. The moss beds remove sediment from the water and have a small restraining effect on the heavy run-off from summer rains. But it is during the winter months when stream flow from the snow-fields is small and the requirement for hydro-electric purposes at a maximum, that they perform their most important function. The saturated peat has a higher thermal capacity than the surrounding soils and the moss beds are fed from deep springs where the water is warmer. Thus they hasten the melting of the overlying snow and augment winter stream flow.

Mountain Gentian (*Gentianella diemensis*) growing out of sphagnum moss.

Soldier beetles (*Chauliognathus lugubris*) on flower of the native Yam (*Microseris lanceolata*).

The moss beds are sensitive to trampling, and past grazing and firing led to a reduction in their area through the formation of drainage lines and ultimately gullies.

I found some native Yams, which at first glance look like dandelions but, particularly at altitude, are larger. They can easily be distinguished from dandelions because the florets end in a terminal row of 'teeth'. The reason for the name is a small underground tuber which is claimed to have a coconut flavour. Yams are widespread and formed part of the diet of the Aborigines in south-east Australia. In the Alps it is believed that they were an important food item during the summer when the main reason for the Aborigines' presence was the Bogong Moth feasts.

I noted an alpine cicada perched on the trunk of a snow gum. This insect is a relic of prehistoric times, with only one living relative, a closely related species in Tasmania. Fossils from the northern hemisphere attest to their antiquity. Relict insects are sometimes found in so called 'adverse' environments which have remained stable over long periods of time, such as this one.

Although both male and female produce no audible sounds it has been found that they do have tymbals and communicate, but outside the range of human hearing. So far the receiving organs, the tympani, have not been identified. The cicadas are unusual in appearance, being hairy and with a pigmented wing pattern.

Candle Heath (*Richea continentis*) in sphagnum moss bed.

FEBRUARY 25

This morning I took a short walk on the Snakey Plains Track as a preview to the trip I plan to take to Dargal Mountain. From the map there does not appear to be any track to the summit and I will have to make my way directly up from Snakey Plains. I lugged my camera equipment about 2 kilometres through attractive snow gums and Candlebarks. The tree trunks were beginning to show their 'autumn' colouration, bright pink and salmon flashes ranging to deep crimson on some trees. The cause of this change in colour is the same pigments, anthocyanins, that produce red autumn colours in the leaves of deciduous trees.

One assumes that the anthocyanins form in bark about to be shed, in the same way they do in autumn leaves. They are produced in the presence of sugar which is trapped by a sudden cold spell and subjected to intense light. The colours begin to appear at these altitudes as summer wanes and the nights become cold, and are displayed on the warm and sunlit side of the trunk and main branches.

During my walk I saw signs of the old cattle runs or snow leases, including fences with posts now missing and rusty and broken wires. Lichens covering the posts showed that no cattle had rubbed against them for many years.

Bark colours: Mountain Gum (*Eucalyptus dalrympleana*).

Snow gum (*Eucalyptus pauciflora*).

Bark colours: snow gum (*Eucalyptus pauciflora*).

Bark colours: Mountain Gum.

In the days of the mountain cattle, cows and calves and dry cattle were driven up each year in November from the home properties in the river valleys. They spent the summer months on the High Plains, grazing particularly on the native and introduced broad-leaved plants growing between the tussocks, the snowgrass becoming more palatable during flowering in mid-summer. Clover was spread by stockmen over the years and can still be found in places.

It was efficient to run cattle in the mild environment of the High Plains and to spell the home property. In most years there would be sufficient feed in the home paddocks, carried over from spring and summer, and fresh green pick following the autumn break, to last the stock through the tough frosty winter of the valleys.

To manage these snow leases required superb horsemanship and expert knowledge of the locality. Huts were built and maintained on most of the leases and rock salt was widely used as a lure to help contain cattle within the boundaries and to aid in mustering them. By reducing the amount of salt towards the end of the season it was possible to more easily assemble cattle at mustering points as their craving for it increased. Use of wire fences along ridges and natural features such as forests and rocky outcrops also helped to restrain them.

The cattle were usually brought down again—and in fine condition—at the end of April. (Occasionally they were trapped in the mountains by snowstorms before

Verandah of cattleman's hut (Wheelers), near the site of the Toolong Gold Field.

they could be mustered and driven out, and hundreds perished.) In the early days, before controls on fire-lighting were introduced, the high country would often then be burnt to get rid of the old tussocks and scrub and to provide green pick for the next season.

However, an act proclaiming the Kosciusko State Park of 5000 square kilometres in 1944 saw the beginning of the phasing out of grazing. The advent of the Snowy Mountains Scheme and the acceleration of the long-running debate on soil erosion culminated in the exclusion of all grazing from country over the height of 1370 metres in 1958. The park came under the control of the National Parks Service in 1967 and all grazing ceased in 1969.

Although the perceived effects of grazing on the flora and on wildlife generally was a factor, the major reason for its exclusion was the fear that the continued rate of soil removal and deposition in the dams could affect the long-term viability of hydro-electricity and irrigation schemes. Today High Plains grazing remains a controversial issue. In the adjacent Victorian High Plains cattle continue to be grazed in specified sectors, supervised by the Soil Conservation Authority.

There is still debate as to the effect grazing has on the alpine flora, on the rate of soil erosion, the speed of run-off and the quality of the water. Some of the more obvious damage appears to have resulted from periods of very heavy stocking, particularly with sheep, during major drought, as in 1902 and 1914. Fires have also played an important role in the changes that have been recorded, both those lit by graziers and occasional catastrophic wildfires, particularly that of 1939 which swept in from outside this region.

In the Snowy Mountains it was much easier to accept the generally well-reasoned and documented arguments justifying the decision to remove cattle if you were not directly involved in the industry. For those people whose cattle operations were vitally dependent on access to the alpine and subalpine pastures, the decision came as a bitter blow.

On my walk I also noted a few old snow gums with prominent fire scars on them. You don't see many trees with obvious fire damage now. Of course there are fires lower down but it is more difficult to start a fire above 1300 metres and usually requires some element of human encouragement and persistence.

Looking at these fire-damaged trees reminded me of a visit I made to the Snowy Mountains in the late 1950s. I recall being shown a demonstration at Cooma by Snowy Mountains Authority scientists of the changing incidence of fire on the High Plains since white settlement. They had designed an elegant method of demonstrating this—it was simply a cross-section of a snow gum. Snow gums, like Alpine Ash, have distinctive annual growth rings. The sharp change in seasons

Fire scars on snow gums (*Eucalyptus pauciflora*) on Snakey Plains Track.

means a clear difference in density and colour between the light spring and the dark autumn wood. It is thus relatively easy to read the passage of the years on a cross-section of an old snow gum. When a snow gum is burnt in all but a very hot fire which will destroy it, a scar is formed on the surface of the trunk and remains for the life of the tree incorporated in the growth rings. Since the growth rings can be tabulated in years it is possible to determine when fires occurred.

In the case of this tree, which had its genesis about 1750, there was one fire in the first 100 years and thereafter at an increasing rate. The occurrence of fires from about 1880 to the time that grazing ceased was about once every five years.

The object of this exercise was to show that past fires were in fact mostly caused by Europeans and were not due to natural causes such as summer thunderstorms.

It was also contended that the Aborigines burnt the subalpine and alpine regions, but it now seems more likely that they did not light extensive fires in the high country. The High Plains never carried a large native animal population and it is generally believed that Aborigines lit fires mainly for the purpose of catching game and to stimulate new grass which would later be attractive to animals. It appears likely that the bush-fire smoke reported by early travellers in the High Plains was from fires at lower altitudes — in the predominantly dry sclerophyll forests which the Aborigines did burn and which were also prone to lightning strikes.

Resting at a small spring I met up with some Grey Fantails. These are tame little birds and with their lively acrobatic flight they came right up to me to investigate, but I still didn't succeed in getting a photograph of them. Although their actions appeared friendly and curious, they may have only been trying to keep my interest away from their nests — although it would seem late for them still to be nesting.

Female sawfly (*Pseudoperga sp.*) protecting young larvae.

I was pleased to find several examples of the female of a sawfly species (*Pseudoperga*) tending their young while on the leaf. These insects exhibit a very unusual maternal instinct towards their progeny. The females were most attentive to the young larvae, sitting over them like broody hens, and would not readily be removed from their presence. Once the larvae reach about a centimetre, however, the females lose interest and you see larger larvae fending for themselves.

Later, as I was coming up the drive to the homestead, there was a long line of sheep interspersed with cockatoos, feeding on some oats. I stole up sufficiently close to get a good photograph. Photographing Sulphur-crested Cockatoos is not easy as they have a strategy of employing a sentinel bird, which roosts in a nearby tree and warns the others of impending danger. It is hard to 'beat the system', unless the cockatoos are distracted as they were today by the prospect of a huge feed of grain.

Although they have always been common, the Sulphur-crested Cockatoo is an example of a bird whose numbers have apparently been increased by the clearing of the original woodlands and forests for the development of agriculture, and the consequent increase in seeds, roots and berries suitable to their diet. They are also partial to thistles and the seed heads sustain them when they can't get sufficient grain. Normally in grazing districts they do not cause much concern, but when an early oat crop is sown they will pull up newly emerged plants. During the last drought we were feeding quite a sizeable portion of our precious grain reserves to the Sulphur-crested Cockatoos, because there was no way of preventing them joining in with the sheep and cattle.

Sulphur-crested Cockatoos and lone magpie.

FEBRUARY 26

This morning I went up to our largest dam, covering about 1 hectare, and built during the 1967 drought in case the creeks ran dry. It hasn't served any real purpose except briefly in the drought of 1982–83 when it serviced the failing stock water reticulation system, but it is an excellent location for birds. I saw a good number today, including King Parrots, Red-rumped Parrots, Sulphur-crested Cockatoos, Galahs and Crimson Rosellas. In addition, the Red-browed Finch and several smaller birds including Superb Blue Wrens and Grey Fantails appear to be nesting in the area. Among the water birds were many Wood Ducks, Blue-winged Shovelers and Black-fronted Dotterels.

The insect life on the dam included many sleekly streamlined dragon flies. The most ornate was a large turquoise species about 6 centimetres across the wing.

OPPOSITE PAGE:
Cloud reflections in farm dam.

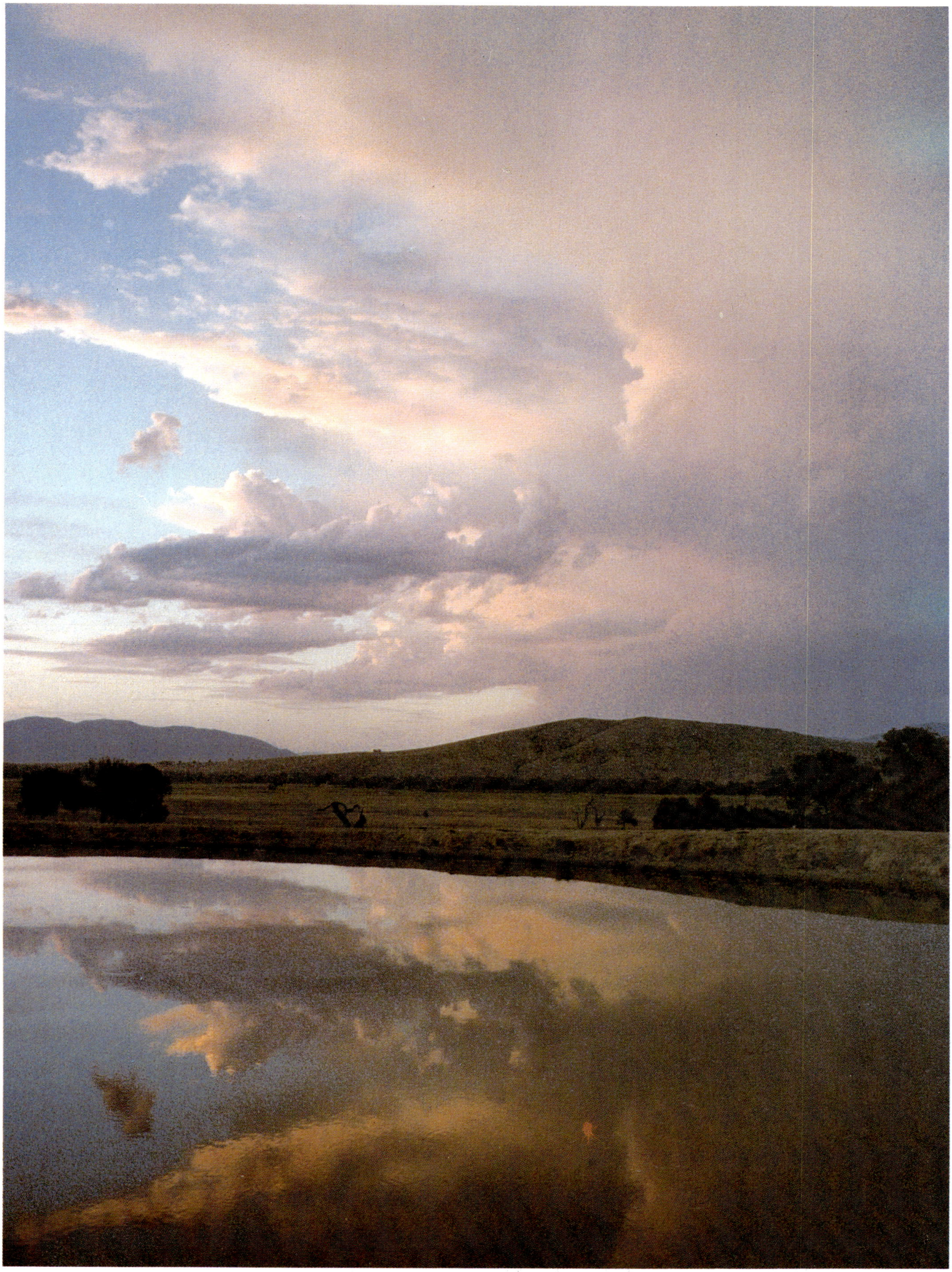

King Parrot.

Dragon fly (unidentified).

The large compound eyes of a dragon fly (unidentified).

The warning 'eyes' of the hawk moth caterpillar (*Hippotion celerio*).

They skimmed rapidly across the surface of the dam, occasionally seeming to caress the surface with the tip of the abdomen, seeking out insect prey which they seize and clutch in their long spiny legs. Then one would unexpectedly alight on a rush and provide an opportunity for a picture and I spent a frustrating period trying to photograph them. By the time I had my lens in focus some slight movement would attract the dragon fly's attention and in a split second it would be gone.

The secret of the dragon fly's reaction to a moving object lies in its large compound eyes. Compared to the size of its body the eyes are enormous. Even without moving its head the insect has an almost complete 360° range of vision. The compound eye of a dragon fly may have 25 000 or more individual hexagonal facets. Each represents the corneal lens of a tiny optical system complete with its own light-transmitting cone and retinal cells, and isolated from its neighbours by a layer of pigment cells.

Compound eyes are very sensitive to movement. It is generally believed (though seemingly impossible to prove) that the image the insect receives is made up of a mosaic of a huge number of minute pictures, each a part of the whole scene. Another dragon fly in flight or a swooping bird would appear as a succession of images across the facets.

The image received from the dragon fly's eye cannot be nearly as sharp as that from the human eye since it does not focus, and it is probably unclear beyond a metre or so.

At the edge of the dam I looked in vain for the early stages of the life cycle of the dragon fly. The immature or nymphal stage is carnivorous, living on other small aquatic insects. Frequently you will see the nymphal shells clinging to rushes, still looking very lifelike.

Late summer afternoon, farm dam.

On my return to the cottage I discovered a caterpillar of a hawk moth (*Hippotion celerio*). This caterpillar can effect a startling change in its appearance. Its first line of defence is its excellent camouflage, though on the smooth, regular jarrah flooring of the verandah it stood out very clearly. Placed back on the trunk of the large vine from which it had wandered it merged in well with the background. As soon as I touched the horn-like tail, huge eyes unfolded on either side of the head. Of course these are not real eyes, but they give that appearance. This dual defence mechanism is common among insects. They depend first on camouflage and should that fail they resort to some form of scare tactic.

I have seen these hawk moths feeding at jasmine flowers on the verandah on a warm September evening. They have the uncanny ability to hover motionless in front of the flower while feeding, their long proboscises lowered into the corolla extracting nectar for quite a few seconds.

MARCH 10

Before first light this morning I left for Farrans Lookout, a vantage point above the Murray close to where it is joined by the Tooma. On the way I stopped at the gap on the ridge separating the valleys of these two rivers. Stretched out below was a perfect sea of white mist. It blanketed the plains of Towong Hill and Khancoban stations and beyond the farthest reaches of the Murray valley. The mountain tops stood out like islands and the pre-dawn light cast a pale pinkish hue on the scene.

These early morning mists are a feature of the autumn here. Usually they don't appear until later in the season, when the earth cools sufficiently to allow the water vapour in the air to condense on tiny nuclei such as dust particles, but this year has been colder and there has been so much rain and moisture in the air that the mists have come in earlier.

Autumn sunrise, Tooma valley.

To my mind the mist is very much like a body of water, even having its own high tide mark. Before dawn each morning it settles at almost the same level on the hillsides bordering the valleys. It seems to sense the approach of dawn and starts to move. The pearl-white flatness of the cloud surface curls upwards in a series of wisps and within minutes you are enveloped in fog. Just before this happens there are great opportunities for the photographer. The first rays of the rising sun cut through the dispersing cloud and cast an ethereal light over what would normally be quite ordinary objects.

The view from Farrans Lookout is perhaps the best known of this part of Australia. This morning my eyes could follow the course of the Murray for several kilometres across its wide flood plain, and the main range stood out starkly against the skyline.

Dawn rays of sun filtering through rolling mist near Farrans Lookout.

Less than a kilometre up-river from the lookout is the site of the old 'Lighthouse' crossing. Before the bridge was built at Towong in 1939 this was an important crossing point on the Murray. In the early days people swam their horses across, or relied on a boat, and the Whitehead family at the old Lighthouse homestead high above the crossing constantly helped travellers negotiate this treacherous stretch of river. Nine people are known to have been drowned during the ninety or so

Early morning view from Farrans Lookout, south-east towards Mount Kosciusko and the Murray River.

OPPOSITE PAGE:
Sunset silhouette.

Galahs.

years it was used, in most instances failing to make a sharp turn with their horses in midstream to follow a gravel bank and to avoid a whirlpool immediately downstream from the crossing.

On the farm the season still continues in great style. True, there is some evidence of drying out, since we have not received much rain in the last two weeks. In this area the soils derived from granite have a low water-holding capacity. During the warmer months pastures respond very quickly to heavy falls of rain, but wilt equally as fast if they don't receive follow-up showers.

The transformation wrought by the seasons is dramatic. A bountiful and seemingly endless spring can change within a week at the end of November as the temperatures rise and the rainfall diminishes. The hills change from light green to brown almost overnight. However, even in high summer there can be a sharp reaction to any summer rains on the fertile, well-structured soils of the river flats.

This afternoon we went down to the Murray River at Bringenbrong (said to be Aboriginal for red kangaroo or red ant hill). It is an old station, although considerably smaller than it was in the 1850s, and lies between Towong Hill and Khancoban stations on some of the richest of the Murray River flats.

The children had a great time swimming in the river, jumping off one willow bough, being swept downstream and catching another 50 metres further on.

There have always been Murray Cod in the river, even up this far, and you still hear stories of the big ones caught and those that got away. Although of course not given anything like the publicity of our marsupials, the Murray Cod and the other indigenous species of the Murray (including Macquarie Perch and Trout Cod) are unique among the world's fishes. They have evolved in isolation over a very long period of time, adapting to the fickle character of the river with its wide fluctuations in water level and temperature.

Generally, however, native fish have not fared well since settlement. The clearing of streamside vegetation, the removal of snags, siltation and pollution have all probably had some influence on their numbers and in more recent times the construction of dams to control stream flow has tended to reverse the natural fluctuation in flow with major water releases in late summer and autumn. Flooding, with its effect on the nutrient status of the water, has been greatly reduced and water released at the base of the dam walls yields unseasonably cool water in summer, unfavourable for the breeding of native fish. On the other hand, important introduced fish have benefited from some of these changes and trout have been gradually extending down the Murray with cooler summer water temperatures.

Along the river today the only native plant species I could see were the old River Red Gums. These would be at least 200 years old and if they had memories would recall that they were once unexceptional members of a large open woodland. Today they stand out in dramatic isolation on the river flats, surrounded by willows lining the stream and rows of elms and poplars along the roadsides and the tracks. In just 140 years virtually every native plant has been removed from the flood plain of the Murray.

Yet on the rocky hillsides above and no more than 3 kilometres distant are places where you would find many native plants and few weeds. The stony soil held no appeal for the early settlers and attracts no attention today because it has little agricultural value.

Just before sunset I came across a group of noisy Galahs in a tree. I don't think we fully appreciate how beautiful these birds are. If they were rare they would be regarded as something very special, but we tend hardly to notice them. Eventually they took off in a great squawking group to quickly assume their characteristic pattern of flight. They are most graceful, streamlined birds in the air. The flock wheeled and dived, alternately displaying pearl grey top and rose pink underside, all in perfect synchronisation. Yet on the ground they are just dumpy little parrots, strutting around, fighting noisily over grass seeds and other morsels.

The Galah is another bird which has benefited by man's clearance of the forest. They have spread around the coastal areas in the last fifty years, and here the opening up of the wide river valleys and the introduction of crops, improved pastures and the proliferation of thistles has provided them with more food.

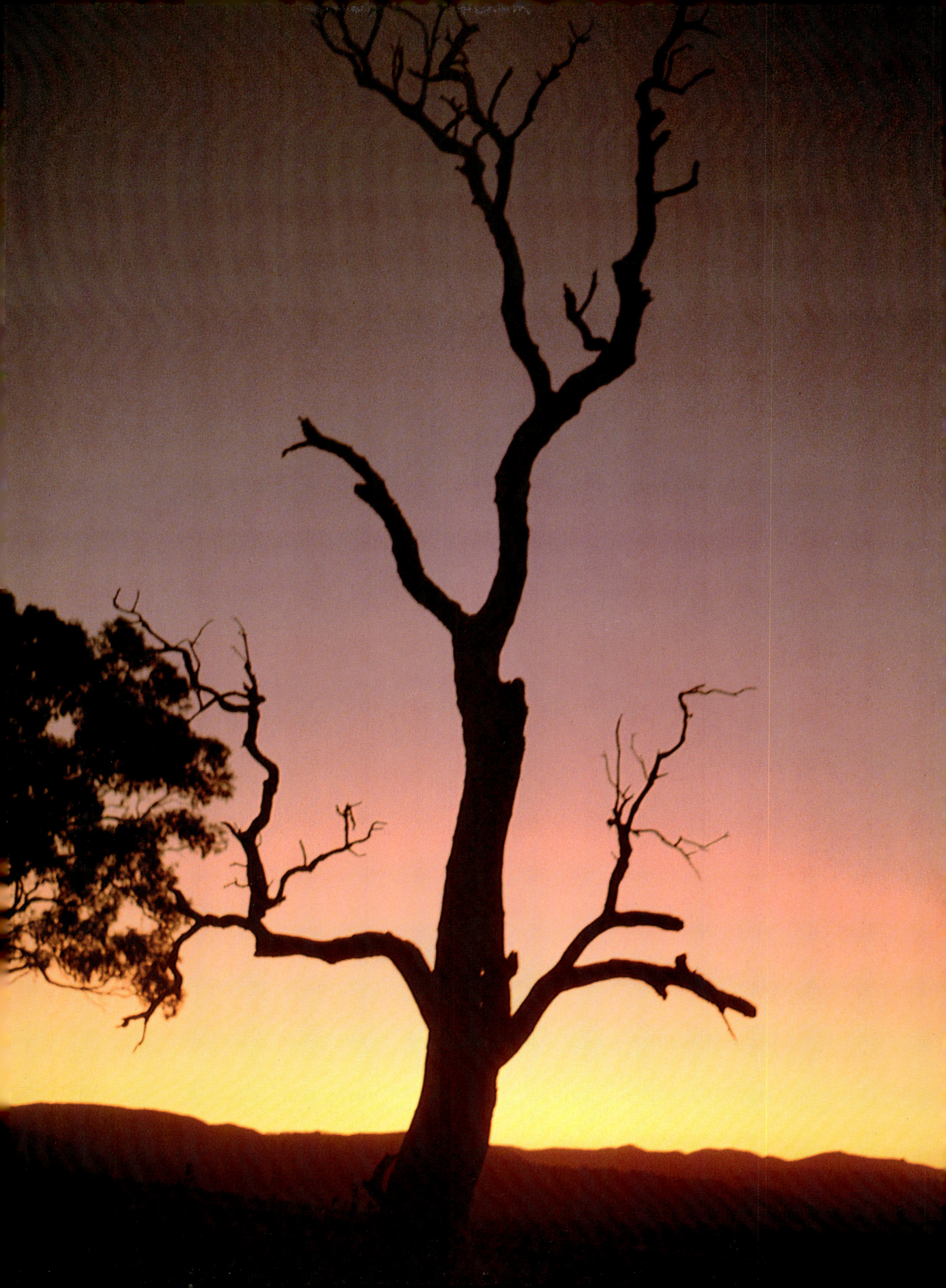

Diamond Weevil (*Chrysolopus spectabilis*).

March 17

I went out at dawn this morning to Farrans Lookout and saw a Brown Hawk chasing a flock of Red-rumped Parrots. It swooped suddenly and plucked one from the air, wheeled and flew off to a nearby tree. The hawk surprised me with its speed and manoeuvrability; I wouldn't have thought it able to catch such a swift bird in flight. It all happened so suddenly (too quickly for my camera) and was yet another reminder of the ever-present 'law of the jungle'.

I visited a small patch of bush quite near the cottage—one of those relic areas, probably much the same as it was over 100 years ago, and not much changed by the grazing of cattle or sheep. I was delighted to find a Diamond Weevil perched openly on a Silver Wattle. The slow moving weevils are the most prolific family of all the world's creatures and there are reckoned to be perhaps 10 000 species in Australia, 5000 having been named. Weevils, sometimes called elephant beetles because of their long snout or rostrum, survive and thrive in their millions, protected by camouflage and by their very hard cuticle or outside covering.

The camouflage of the Diamond Weevil seems contradictory. Sit it on the back of your hand and it is a brilliant insect, about 2 centimetres long. Its basic colouration is black with a series of indentations on the surface of the wing cases, each patterned with a silvery-green pigment. Place it back on the feathery leaves of the Silver Wattle and it is easy to see at all. The outline of the beetle is broken up and the green markings are well matched with the colour of the wattle.

Common Grass Blue butterfly (*Zizina labradus labradus*) opening its wings at dawn to catch some warmth from the rising sun.

Leaves of Alpine Ash (*Eucalyptus delegatensis*).

March 18

I left early yesterday morning for my trip to Dargal Mountain and was on the Snakey Plains Track before 8 o'clock. The rays of the sun were just appearing through the snow gums, picking up the colour of the fallen bark, silver on top and salmon pink on the underside. I was pleased to find that the everlastings were living up to their name and were still out in profusion.

The track was easy for the first 3 kilometres, but the climb to the ridge above Wolseleys Gap was fairly solid. I guess it's about a rise of 300 metres and it took a little over half an hour. I walked at first through the Alpine Ash and Mountain Hickory Wattle, the sun streaming through bright green and red tipped translucent foliage, a faint smell of eucalyptus vapour in the air, to finally reach the stunted and twisted snow gums on the crest. This is a hungry-looking exposed ridge with deformed gums that have obviously endured the cold, hard winds of winter and on a harsh stony soil that does not allow them to reach their potential stature.

The walk down to Snakey Plains was easy, with the ground soft underfoot. Snakey Plains is a small grassy meadow on either side of a stream, interspersed with patches of low scrub and moss beds. One imagines its name stems from the occurrence of Copperheads, the common snake at this elevation. I didn't see any, although I must say I didn't look too hard.

I found a black caterpillar (Anthelidae), copiously covered with hairs, walking along an old whitened snow gum log. It made no effort to conceal itself, but after I photographed it I touched it and it sprang immediately into a tight ball and remained thus for the next ten minutes. After a while it furtively lifted its head, decided all was safe and resumed its slow and measured walk across the log. This defensive ploy is quite common with some caterpillars and a number of adult moths will also feign death when you disturb them.

'Woolly bear' caterpillar (Anthelidae) coiled up defensively.

I could see Dargal Mountain from Snakey Plains and still hoped there might be a track. This hope was soon shattered and it was clear that it was going to be a matter of heading directly at the mountain, which I reckoned to be only about 2 kilometres from where I stood.

The first part proved fairly easy. I walked through the snowgrass to the edge of the plain and crossed one or two creeks to a small sphagnum moss bed. Shortly after this the going changed dramatically. I was now in dense shoulder-high bush and from here on in it was a 'scrub bash' all the way. There was one brief interlude when I passed around a moss bed and I took time out to photograph a Mountain Gentian, which is superficially like a white buttercup but with dark purplish-black stripes. Its appearance was enhanced by the presence of a bright orange-bodied native bee feeding on the nectar within the flower.

Native bee (*Exoneura sp.*) on flower of Mountain Gentian (*Gentianella diemensis*).

It took me about two and a half hours to climb the mountain. I have had a fair amount of experience at pushing through scrub and normally you look for and find a break here and there, but there were few breaks on this climb. The snow gum trunks were very crowded, presumably the result of regrowth after fires. The pack I was carrying, which weighed about 20 kilograms, proved a liability. The trees were so close that frequently it wouldn't fit between them, and at times it took me as long as ten minutes to cover a hundred metres. With some relief I finally saw a metal trig point marking the summit. I expected to see some means—a track or trail—along which the materials for the trig point were carted up to this wild, remote peak. However, I couldn't find any way that this could have been achieved other than by lugging each individual piece of metal up through the scrub.

At the summit I took off my pack and sat down to contemplate the scene. The view back towards the Murray valley was covered in a haze of smoke. Farmers anxious to prevent a repetition of last year's bush fires were burning off before the winter. Looking south towards the Kosciusko massif and east to Jagungal there were clear views. My contemplation of this vast panorama was interrupted, painfully, by the concerted attack of jumper ants. These yellow-fanged monsters are closely related to the bull ant and, like the bull ant, seem to react to an alarm system. I have been bitten on many occasions in the bush, usually after accidentally standing on a nest. The ants seem to have an arrangement whereby at the signal of a leader they all bite simultaneously. This can have a dramatic effect if it happens on some of your nether regions. In this case they struck me through the backside of my jeans—so they obviously have powerful stings. For this and other reasons I decided to move to a seemingly better site about a hundred metres distant. In fact, that hundred metres took me a further ten minutes, through very thick scrub on the top ridge.

A banded native bee (*Amegilla sp.*) asleep, speckled with morning dew, its jaws gripped tightly on a grass stem.

I went back for my pack and, on returning to this new camp site, stumbled across an unusual fungus. It is sometimes called the Starfish Fungus (*Aseroe rubra*). It pops up dramatically from an egg-like sac lying just below the surface of the soil. The mature fruiting body looks like some unworldly form of sea anemone. Several bright red Medusa-like arms radiate out some 5 centimetres from a central void, each of the arms sharply forked at the end.

It has a powerful and fetid odour which attracts flies. This is the fungus's mechanism for distributing its spores, which lie in a glutinous mass on the surface. The flies pick up the spores on their forelegs and mouth parts and transport them elsewhere. There were perhaps as many as twenty flies on this particular specimen.

I can't say I had much joy at my new camp site either. Instead of jumper ants I now had small black ants, in what seemed like millions. It was not a matter of locating a camp site but of finding somewhere to actually stand without being covered by ants. I have never experienced anything quite like this. Fortunately the rays of the late afternoon sun soon began to fade, and the ants lost their frenetic energy and the situation improved considerably.

The Starfish Fungus (*Aseroe rubra*).

During the evening it turned very cold. I noticed many moths silhouetted against the drifting clouds. The moths, almost certainly Bogongs, were flying rapidly in no particular pattern. Bogong Moths are very much a part of the history, and the natural history, of the south-eastern highlands of Australia. They breed on medics and Capeweed in grasslands, in the drier climate west of the mountains. Bogongs migrate to the mountain tops early in summer and return to the breeding ground in late March. Entomologists call this behaviour facultative diapause. By retreating to a milder climate the moths avoid the dry adverse environment of their breeding ground during summer. On the mountain tops the Bogongs frequent caves and cracks in the rocks. They remain essentially immobile during this period, leaning over each other in a formation which duplicates the image of the tiled pattern formed by the scales of their own wings. One cave may contain many thousand moths. Towards the end of diapause they begin to fly again around the mountain peaks in the evening and return to their breeding grounds which may be hundreds of kilometres to the north or west.

This pattern of behaviour must have been going on for a great length of time because nematode parasites have become specifically adapted to the lifestyle of the Bogongs. The nematodes remain in the caves during the winter and reinfest the returning Bogongs during the late spring and summer.

The Bogongs were regarded as a delicacy by the Aborigines. They smoked them out of crevices between the rocks, collecting vast numbers in bags made from the fibres of Kurrajongs and rice-flowers, and formed a paste of their bodies which they roasted in hot ashes. The bodies of the moths comprise more than 50 per cent fat. Early explorers and naturalists who tried the fare differed in their assessment. Some said it was very tasty with a nutty flavour and others that it was quite inedible. The moths must have been beneficial, for the Aborigines were reported to have returned from the mountains to their tribal grounds below in great physical shape.

The supply of Bogongs was apparently sufficient to support a large itinerant population of several mountain tribes and the feasts appear to have been a focal point for inter-tribal ceremonial activities as well as a valuable summer food source. So much was this annual feasting a part of local Aboriginal culture that the name for a mountain peak, in the dialect of the Aboriginal inhabitants of nearby areas, was the word 'Bogong' and Mount Bogong, Victoria's highest peak, carries that Aboriginal name.

MARCH 19

I was woken early in the morning by a flock of young Crimson Rosellas in the tree above my head, then a busy group of Striated Thornbills arrived. I soon discovered that the whole peak was enveloped in thick grey fog, and camping immediately above the western scarp I was looking down on a great grey thousand-metre void.

To the north the snow gums, silhouetted against a ridge, appeared ghostly in the fog. The infinite variety of shapes this tree can assume and still retain its essential grace is one of the indelible memories of any visit to the Alps. It is the weight of the snow during winter that creates the beautiful arching effects of the branches.

The parrots left and were shortly replaced by very raucous black cockatoos. I don't think they saw me and they chattered noisily and unconcernedly among themselves, allowing me the opportunity to photograph them. I was about to take a second photograph when I heard a rustling sound coming from the edge of the rocks and I turned to see a wild dog about 10 metres away. When I investigated more closely I saw that it was not a pure Dingo but looked more like a Blue Heeler, though it turned out to have the characteristic howl of a Dingo. To my surprise it was quite unafraid and stood there gazing calmly at me—a look of pure curiosity. I suppose we stared at each other for about a minute and then I decided that perhaps such a large dog could become a nuisance, particularly as I eyed my provisions for breakfast. So I bade it be off and it turned slowly and wandered out of sight. I could hear its howls for perhaps ten minutes afterwards. I am sure that this animal had never seen a man before.

Ten kilometres from here many wild dogs roam the borders of the National Park and the farmland. They are extremely shy and cautious and would certainly not allow you to approach so closely.

I must say my experiences on the summit of Dargal Mountain made a strong impression on me—the harsh screeching of the black cockatoos, the mournful howls of the Dingo and the eerie background of ghostly snow gums. I also wondered just how I was going to get down if the fog held all day.

I left the top at about 8.30 in the morning and made my way very gingerly down the spur. I was concerned that I might wander off it in the fog but was fortunate that I was heading due east and could set my course on the lightest portion of the sky. I did experience a few anxious moments but finally emerged an hour and a half later—about 200 metres from where I began my ascent from the plain yesterday.

I was very relieved about a quarter of an hour later to find myself back on the Snakey Plains Track. I estimated that I'd have about an hour and a quarter's walk back to the car, but I'd hardly started when a heavy downpour began. It continued for the whole of my walk and became heavier as I progressed. The rain was accompanied by much thunder and lightning and some hail and I counted the time between the streaks of lightning and the peals of thunder. I took a sort of morbid interest in how close the flashes of lightning were—according to my calculation no further than 400 metres, based on the fact that sound travels at 400 metres per second. I was drenched and walked the last section in water up to my ankles.

These mountain storms give you some idea of the erosive power of water in this terrain. Water was cascading down the steep parts of the track and off the side into the tussocks of snowgrass. It is easy to visualise, with reduced vegetation, how great the damage caused by erosion can become. One of the problems is that

Snow gums (*Eucalyptus pauciflora*) in early morning fog, Dargal Mountain.

Bark of snow gum (*Eucalyptus pauciflora*).

vegetation at this altitude, once removed, takes many years to come back. On some very exposed sites it may not come back at all.

The instability of the surface soil makes germination of new seedlings difficult. A process called frost heaving causes the upper layer of the alpine soils to expand during the evening: water within the soil pores turns to ice and then contracts again during the day as the sun warms the surface and melts the ice.

March 20

I visited some old red gums on the property today. I don't know how old they are but from their size I would estimate that they would be in excess of 200 years. Each year some of these ancient red gums die. We have planted new seedlings in several paddocks, but these grow slowly and are difficult to nurse through the early years. Red gum seedlings grow most readily on the flats but they are very fragile on the rising country. Plant a young one near an old flourishing tree on the hills and it will almost certainly die. It seems that only in very wet summers, when the free draining nature of the soil is offset, do the red gums strike naturally and persist on this harsher, lighter ground.

Old red gum (*Eucalyptus blakelyi*).

The mature red gum, unlike many other paddock eucalypts, is resistant to the grazing animal. However, cattle and sheep destroy the young seedlings. The way to counter this is to plant new ones, each with their own tree guards, or fence in an area to the lee of an old tree. Seeds will be distributed by the prevailing wind and some of these will grow and flourish within the enclosure. Unfortunately this is both costly and a potential harbour for rabbits.

I had a look at the forest which borders the transmission line behind the property. The Snowy Mountains Authority keeps the strip of about 150 metres in width fairly clear of undergrowth and it is covered mainly by Kangaroo Grass, some wattles, dogwoods and a few Bursarias. I found one Bursaria in flower all on its own, perhaps the last one flowering in quite a large area. It is a most attractive plant to insects and emits a powerful, almost sickly scent from its profusion of small white flowers.

Among many different species of insects on this plant was a jewel beetle (*Stigmodera*). Australia boasts some of the most spectacular jewel beetles in the world. This one had a dark metallic black head and thorax, with deeply etched brick-red wing cases.

This beetle bears a close resemblance to a group of unrelated and unpalatable soft-bodied beetles (*Metriorrhynchus*) which it mimics. Mimicry is one of the more fascinating aspects of insect ecology. It does not seem to be well developed in Australia and is encountered much more often in the tropics where the variety of insects is much greater and, presumably, the fight for survival keener. Some insects have evolved to resemble species which are unpleasant in taste to predators to gain a protective advantage. Mimicry may extend to imitation of quite unrelated species. Insects as diverse as moths, bugs, wasps and flies may all be found imitating some distasteful beetle. So subtle and faithful are the adaptations that when the mimicked species or model develops geographical races, mimics take on the pattern and colouration of the particular race in that locality.

Later this evening I encountered an even more unusual example of mimicry. A fierce-looking wasp I caught on the Bursaria in a bottle to examine at my leisure proved on close examination to be none other than a longicorn beetle (*Hesthesis*). The wing covers in this genus have shrunk to mere buds, exposing transparent wings like those of a wasp, and the fully exposed abdomen is decorated with typical yellow wasp-like bands to complete the illusion. The behaviour and movement of the beetle has the same restless, furtive quality you associate with wasps.

Soft-bodied beetle (*Metriorrhynchus sp.*): the model.

Jewel beetle (*Stigmodera sp.*): the mimic.

Longicorn beetle (*Hesthesis sp.*) which imitates a wasp (*Pseudozethus sp.*)

APRIL 15

Today I visited an area on the Kiandra-Khancoban road at an elevation of about 1350 metres. The country here has now taken on the appearance and feeling of autumn. Temperatures are cooler and there are fewer birds and flowers, and the insects are more sleepy than a month ago. I spent some time photographing snow gums and Mountain Gums. They are both white-trunked trees and assume similar bark colourations during autumn.

I found a female Mountain Grasshopper on a snowgrass tussock. The stout body is basically black and ringed with enamelled red and blue bands, and it looks for all the world like a large old-fashioned boiled sweet. When the abdomen is covered by the dark crinkled leaf-like wing covers you see no evidence of colour, but the sudden movement of my hand caused it to raise its wing covers vertically, which exposed the brilliant markings.

There are quite a few signs of rabbits at this height in the National Park and this surprised me. One doesn't expect to see rabbits at this altitude but these have managed to adapt to the cold conditions.

At a lower altitude (700 metres) I came across signs of extensive digging around the roots of tea-trees. It didn't look like the work of wombats and could well be wild

Female Mountain Grasshopper (*Acripeza reticulata*).

Bolete fungus.

pigs, which have extended into the Kosciusko National Park in recent years. The wild pig poses a serious potential threat should an outbreak of foot and mouth occur here. It would be extremely difficult to eradicate this disease if the wild pig population became infected. Wild pigs have not as yet caused us any trouble but in other areas they damage crops and pastures and take small lambs.

Rabbits, the increasing number of pigs, and wild dogs, point to some big management problems for the limited manpower available to the park authorities and the difficulties are compounded because the park is surrounded by farmland.

I saw several Swamp Wallabies on the road during the evening. They are relatively tame and you can get quite close to them in a car. They don't move nearly as quickly as the Eastern Grey Kangaroos and when they do hop it is a rather tentative jump. They seem more prolific at about 450 to 600 metres, presumably finding the winter conditions at higher altitudes too severe. Also the lower forests, particularly the Narrow-leaved Peppermint, provides the type of dense understorey in which they like to hide during the day.

The Swamp Wallaby is apparently quite genetically distinct from other wallabies and is the sole representative of the genus *Wallabia*. It also differs in its behaviour and reproductive characteristics and retains its fourth molar teeth which enable it to eat tough fibrous vegetable matter, including bracken.

One animal found in the cooler climate of the higher altitudes is the wombat. It seems able to cope quite well at 1200 metres probably because of the relative warmth of its burrows. There were many burrows, some so large I could have almost got into them. It has been found that some can go underground for 20 metres or more. The wombat uses several: the main tunnels have a number of entrances and usually more than one 'bedding chamber', and the smaller burrows are used as temporary shelters.

I have not seen any wombats this year. They are usually nocturnal and when I have seen them during the day they have been affected by a type of sarcoptic mange, and I've wondered whether this influences their capacity to distinguish between day and night. They also apparently take advantage of the weak winter sunshine at or near the snow-line and are sometimes seen grazing on the snowgrass during daylight hours.

The wombat is the largest burrowing native animal in Australia. It seldom exceeds 40 kilograms in weight but has immense strength, aided I guess by its low centre of gravity. As well as digging very substantial caverns, it is capable of moving quite big obstacles. It can be a nuisance to farmers because there is no type of conventional fence that can keep it out, although developments in electric fencing are likely to eventually solve the problem.

APRIL 16

I spent this morning on the farm. It is much greener than on my previous visit. Although the weather has been fairly dry of late, autumn is definitely here and the leaves of the deciduous trees are now beginning their annual display of colour.

The autumn rains we have had will enable us to sow down a hill paddock that failed in the drought of 1982. Last year we couldn't get on the country because it was just too wet. Sowing pastures can be a risky enterprise with costs for seed and superphosphate at well over $50 per hectare.

We are drenching the heifers for fluke and worms at the moment. Liverfluke is prevalent here and, if uncontrolled, sheep can die in large numbers. The symptoms are often not obvious in cattle but a heavy infestation can affect growth in young cattle and milk production in the cow.

The fluke has a complex life cycle involving two hosts. The eggs eventually leave the flukes which accumulate in the bile duct of cattle and pass through the dung. During warm moist periods in spring larvae move out and infect a small snail. They develop inside the snail and in early summer the young fluke (cecaria) are eliminated and move onto the grass. These microscopic creatures are ingested by sheep and cattle usually in the summer. They invade the liver and develop into liverfluke.

In theory you could contain the disease by eradicating the snail from swampy and low-lying paddocks by treating these areas with copper sulphate. In practice this is all but impossible in rough and broken country such as occurs here. The major means of control is by drenching with chemicals.

I was hoping I might find some autumn fungi today but the dry condition of the soil at present precludes this possibility. You can see where the toadstools have been but they have shrivelled up and died. I decided to try instead the wet gullies nearer Khancoban, where the rainfall is higher. While I was there it rained and this made observation difficult, though I did find some fungi.

One of these was a bolete, a well-known group which occurs world-wide. They have the general appearance of a mushroom with stem and cap, but with a major difference. Mushrooms and toadstools, which are agarics, have gills on which spores are formed and deposited on the ground or blown away in the wind. The boletes have no gills, but instead a layer of pores beneath the cap containing the spores. This species had unusually contrasting colours, shiny crimson on top, with a yellow undersurface. The stem was reddish.

Boletes are a somewhat controversial fungi when it comes to their gastronomic possibilities. They have been eaten over the centuries and some have a reputation for being poisonous—one rejoicing in the Latin name *Boletus satanas*. A 19th century writer claimed that it proved so 'devilishly poisonous' that even its emanations caused him to be ill when he was describing it! I believe, if it is properly cooked, this particular species is a treat. Another species, the Cep (*Boletus edulis*), is renowned in France as cèpe, a table delicacy.

I also found a species of the *Pleurotus* genus. I thought it was the luminous fungus and took some home, but tonight when I went out eagerly to examine it there was no sign of any light being emitted and I obviously had the wrong species.

Full moon before dawn.

Flower of Silver Banksia (*Banksia marginata*).

Flower detail, Silver Banksia.

Pleurotus are often large toadstools, found growing as groups on rotting wood. The luminous fungus (*Pleurotus nidiformis*) measures up to 25 centimetres across the fruiting body. I would expect to find it here and have been looking for it for the last few weeks, but I am now pessimistic about my chances as it is probably too cold.

The brightness of these toadstools at night is astonishing and in pitch darkness fresh specimens held close to a page enable the reading of average size print! I remember vividly the first time I found a ring of these green glowing toadstools while walking at night, and can think of few encounters in the bush that are quite as startling. The chemical reaction that produces the glow is well understood but so far no evidence has been advanced to show that it has any value to the fungus.

On my way home I reflected on the attitudes people have towards fungi generally. I've always found fungi interesting, but there is a negative reaction to them which I think stems from the fact that the more obvious ones, like toadstools, are thought to be universally poisonous and others are often associated with plant diseases and, in some cases, diseases of animals and even humans. But in fact fungi are as vital to life as green plants. If fungi and associated bacteria ceased to break down organic matter in the environment, in a relatively short period of time we would be overwhelmed by a huge surplus of undesirable plant debris, with all sorts of consequences which would ultimately make life impossible. In some instances fungi are parasites, but mostly they are acting as saprophytes in forests—living on dead matter—and perform a vital function in breaking down organic matter and releasing nutrients into the soil.

An interesting aspect of many fungi is their world-wide distribution. Their spores are extremely small and light and probably because of this fact they have been transported around the world. In early balloon experiments spores were discovered up to 21 000 metres above the earth and were collected and found to be viable. Such tiny particles can be transported by winds in the stratosphere from one continent to another.

As well, of course, there is the influence of man. Soil and plant material, often containing resistant spores of fungi, have been transported from continent to continent for centuries, sometimes with disastrous consequences. Dieback disease, which threatens vast areas of forest in Western Australia and south-eastern Australia, is caused by a fungus (*Phytopthora cinnamomi*) and is generally agreed to have come from the so called 'Spice Islands' of Indonesia, which were linked by various trade routes to Europe, Africa and North and South America as early as the 17th century. Its entry into Australia remains obscure, although it seems to have first reached Western Australia at the beginning of this century on plants brought in from south-eastern Australia.

While looking for fungi I was struck by the uniformity of the vegetation in the area. This was particularly so with the dominant species—Narrow-leaved Peppermint. Often this peppermint grows in conjunction with other species, particularly Red Stringybark, Candlebark and Manna Gum, but in this case it was virtually a single species stand. It forms an open canopy which can be from 30 to 40 metres high, and you can almost guarantee it will only occur on better watered sites and on deeper well-drained soils. In lower rainfall areas Narrow-leaved Peppermint will be found on a southern aspect and usually along a creek, where moisture is more readily available. Under high rainfall conditions it is frequently associated with an understorey of taller shrubs like musk, Blanket-leaf and pomaderris.

APRIL 17

There was heavy rain overnight, a short sharp burst which gave us around 5 millimetres in about five minutes. It will help the country a little, but our newly sown pasture is on an exposed hillside and this rain may well do more harm than good.

I went out to the Greg Greg Fire Trail some 5 kilometres east of the property. At its outset this trail follows the old stock route along which cattle were driven to the high country. It passes through some surprisingly rocky terrain comprising the gorge of the Welumba Creek.

The soils are mostly shallow, supporting straggly stands of Red Stringybark and in places 'open heath', dominated by Violet Kunzea and Common Fringe-myrtle, no higher than a metre. In places on the slopes, rock comes to the surface, and pockets of poorly drained soil are formed.

I watched an Eastern Spinebill and a Brown-headed Honeyeater working diligently to obtain nectar from the flowers of the Silver Banksia. Banksia is an important component of the bird flora here because it provides a source of nectar in the cooler months when other plants do not flower. Red is usually said to be the dominant bird-attracting flower colour, but today the birds are showing equal interest in the yellow-green banksia flowers.

A high proportion of Australian plant families are adapted to encourage pollination by birds. There is also an unusually large representation of nectar-seeking birds among the Australian avian fauna when compared to other continents, and the co-evolution of birds and plants here has been significant. Honeyeaters are characterised by a brush tongue which they can extend beyond the tip of the bill to reach the source of nectar. Lorikeets have brush tongues, too, and as members of the same family as the parrots have evolved in a parallel form to the honeyeaters.

Bird-pollinated flowers can be classified in an order ranging from the simplest brush-shaped blossoms such as banksia, where the pollen is deposited all over the face of the visiting bird, through tubular types like some species of correa, where pollen is deposited on the base of the beak and the facial feathers, to gullet-type blossoms such as kangaroo paws, where pollen is placed more discretely on the forehead. In another group of flag-shaped flowers (*Brachyseme*) the visiting bird straddles the blossom and pollen is deposited on the abdominal feathers.

APRIL 18

Before first light I climbed to the top of the Lighthouse Mountain. It was misty and hard to find my way. I hoped to beat the rising sun this time and get some pre-dawn photographs from the same point at which I had taken shots earlier in the year. On the way up I was distracted by a group of six kangaroos silhouetted against the skyline. I unpacked my camera gear and set up to try and photograph them but in the end I missed the shot.

Kangaroos and the domestic animals, sheep and cattle, are certainly a lot easier to approach in mist than they are at other times of the day. Perhaps this is because they don't fully appreciate your identity.

When I did get to the top the loss of five minutes was the difference between a good picture and an ordinary one. One of the possibilities I was considering was incorporating the red grass, with its rich russet colour, into some of the scenery. Red grass is a coarse, tussocky perennial species which takes over from more nutritious native grasses. The succession here appears to be that you start with Kangaroo Grass on nearly all classes of cleared country, but under persistent grazing it is replaced, first by wallaby grass and then, particularly on land formerly under red gum and Apple Box, by red grass.

If the fertility is raised by repeated top dressing with superphosphate and the spreading of sub clover seed, annual grasses such as brome and barley grass become dominant. This 'annual' pasture carries much more stock but it becomes bare in late summer and autumn and is less effective than red grass in preventing sheet erosion. The best answer is to sow a perennial grass such as phalaris as well as clover, but this is costly and sometimes difficult on steep slopes.

Red grass is a 'woody' grass and is extremely durable. It becomes unpalatable once it reaches a certain stage of growth and is not eaten by stock, and the tussocks remain through to the next summer. This year the predominance of red grass in the unimproved country is a reflection of the good summer rainfall.

View from the Lighthouse summit, red grass (*Bothriochloa macra*) and grass-tree (*Xanthorrhoea australis*) in foreground.

This morning when I got back to the cottage I noticed that a swift moth had flown in during some light rain we had last night. This is a very large species (*Trictena argentata*), the male with a wing span of 12 centimetres and the female with twice this span.

Dawn from the cottage.

Male of the swift moth (*Trictena argentata*).

Most swift moths are insects of the high rainfall regions but this species is found quite commonly both in coastal and inland Australia. In the more arid regions it lives in association with the red gum. The red gum is an indicator of the moisture content of the subsoil and invariably grows where there is an accumulation of moisture. In dry regions, for example, it is found along creeks or above an underground source of water. Thus the caterpillar of the moth living on the roots of the red gum enjoys an environment consistently high in moisture even in an arid area.

The appearance of these moths often heralds the autumn break. They seem to have the ability to estimate the amount of rain required to ensure that their eggs will hatch and that the caterpillars will have some chance of survival.

Immediately after mating the female lays as many as 50 000 eggs, scattering them at random around the base of the red gum. Huge numbers of tiny larvae must succumb to desiccation as they seek surface roots to feed on and the survivors are then 'sorted out' by the environment and by parasites and predators over the ensuing years of their subterranean existence.

The caterpillar spends several years living on the roots of the tree and before pupation moves to the surface and opens up an exit hole complete with a cap made partly of silk and partly of earth and leaves. The cap when closed is not visible from the surface. The caterpillar pupates at a considerable distance below the outlet. Presumably when the relative humidity in the tunnel reaches a certain level, the pupa works its way back to the surface, pushes the cap up and allows the moth to emerge.

It has been shown that the moth is able to maintain its temperature some 10°C above ambient by the rapid beating of its wings during the one day of its adult life, assisting its activity in the cold periods associated with the autumn rains which bring it out.

MAY 5

I had a look at the lower slopes of the Lighthouse Mountain this morning and was interested in the Kurrajong trees. They are like an exotic tree, perhaps a Camphor Laurel, with shiny olive green leaves, and they grow to about 10 metres. The Kurrajong is a member of the genus *Brachychiton* and as such is a first cousin of the very decorative Illawarra Flame Tree.

In past droughts the Kurrajong was frequently cut down and the foliage used as feed for cattle and sheep. It is not very resistant to fire, and now tends to survive on rocky hillsides, often growing out of fissures in the granite boulders where presumably it is protected from grazing animals. Today I noticed several plants growing next to red gums, just like twin trees. Could it be that the tiny seedling growing up between the debris at the base of the tree and the trunk is afforded some protection?

The Kurrajong was an important item of Aboriginal culture because it was a source of fibre for nets and other purposes.

Burnt grass-trees (*Xanthorrhoea australis*).

A few other remnant species survive on the Lighthouse, despite the fact that it has probably been grazed for about 130 years. The survivors among the rocks include the native bluebell, another blue-flowered plant, Rock Isotome (not now in flower), and a twining plant of the legume family (Glycine). Further up among the rocks where the sheep and cattle have not been able to graze there are a number of small Silver Banksias, a species of pomaderris and Red-stem Wattle.

One of the surprising features of the red gums (*E. blakelyi*) on these stony outcrops is their size and obvious good health. The country is harsh and rocky and so one must assume that the trees tap into substantial underground sources of water, probably held between rock bands at a considerable depth.

Grass-trees have survived, even flourished on the rocky hillsides. Surprisingly they belong to the lily family and their strange primeval shape makes them wonderful subjects for painting and photographing. In this country the grass-tree seems to thrive on the poorest of soil, usually with a northern or westerly aspect.

Grazing does not appear to affect them to any great extent and fire only encourages their flowering, which occurs in spring, occasionally in a spectacular manner, even without the aid of fire. They grow very slowly and it is claimed that some of the larger specimens may well exceed 300 years in age. Although the overall growth of the plant is slow this contrasts with the huge flowering stems which can grow at a rate of 7 centimetres per day in spring. When in flower they attract a great variety of insects and birds, particularly honeyeaters. At night bats visit them, seeking out the moths and other insects clustered on the flower spikes.

For the Aborigines grass-trees were a source of yacca gum (produced by the cortical cells of the leaves) which was used as a type of adhesive. The so called 'death spears' were made by embedding a row of barbs of quartz or other stone in gum along the spear head. The spear shaft itself was often made from the flowering stalk of the grass-tree. It is believed that the Aboriginal tribes inhabiting the south-eastern highlands made particular use of the death spear because of its effectiveness in killing large game, primarily kangaroos, which formed a major part of their diet.

The Aborigines also used the grass-tree leaves in the making of shelters and as a form of bedding. The base of the leaves was eaten and the larvae of longicorn beetles, which bore into the stem of the flowering heads and were regarded as a delicacy.

An extractive industry exploiting the grass-tree developed last century and the gum was used as a basis for varnishes, lacquers and for a variety of other purposes including the manufacture of munitions, citric acid and as a source of alcohol. In 1913 one thousand tonnes of resin were exported from South Australia alone.

Looking east across the valley of Taylors Creek I could pick out a small clump of Black Cypress-pines. They have somehow survived in the gully of a cleared paddock and are remnants of extensive stands, some of which still exist in the adjacent forests along the gorge of the Tooma River. The timber, though more fragile and less durable than the White Cypress-pine (sometimes known as Murray Pine) was well regarded and used frequently in early buildings for flooring and timber joists. Like White Cypress-pine it is more resistant to white ants than the other locally available timber species.

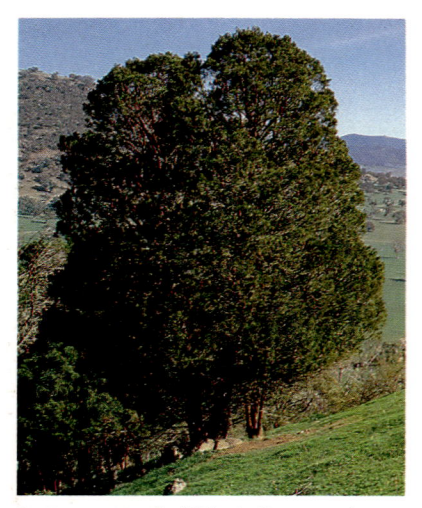

Remnant stand of Black Cypress-pine (*Callitris endlicheri*).

Fern gully in Narrow-leaved Peppermint (*Eucalyptus radiata*) forest; Common Ground Fern (*Culcita dubia*) in foreground, treeferns (*Dicksonia antarctica*) and *Pomaderris sp.* in background.

MAY 6

I had a look this morning at the new sowing in rising country in a paddock known as Mocattas. Mocatta was an earlier owner of the property who cleared some of the timber probably seventy years ago. We are concerned about the development of a few patches of thistle in the pasture. It is always difficult to overcome a weed problem when the emerging pasture plants are young and sensitive to herbicides. Choosing the correct herbicide and the rate of application necessary to kill the thistles when they are growing among clover is a tricky assignment.

On the edge of the new sowing I noted a group of Eastern Grey Kangaroos, which are a feature of this part of the property. On the way back I saw a much larger group, perhaps of twenty or thirty, but they fled into the forest. You don't see much of them during the day when they are resting under the trees and it's usually only in the early hours or more likely the late afternoon that they come out and graze on the edge of the forest. So far they don't seem to have done much damage to the new pasture, probably because the plants are too small and spindly, scarcely reaching over the top of the drill rows.

The Eastern Grey is found all along the east coast, from Queensland to Tasmania, and it merges in western New South Wales with the Western Grey Kangaroo. The two species are closely allied and it is only by blood testing and aspects of reproductive biology that you can separate them. One interesting sidelight is that the Western Grey Kangaroo in Western Australia is relatively resistant to sodium fluoroacetate (1080), which occurs in certain native plants, and to which the species has become acclimatised over a long period.

The kangaroo is an efficient utiliser of forage. Despite poor quality pasture — the basis of its normal food, low in nitrogen phosphate and usually quite high in fibre — the kangaroo is able to convert it to high quality meat. A kangaroo carcass contains about 52 per cent meat compared to that of a sheep with about 27 per cent. In this utilisation of forage the kangaroo is a more efficient animal than either sheep or cattle.

We surprised a Wedge-tailed Eagle as we came around a corner in the Toyota. It was feasting on the carcass of a newly dead lamb some 10 metres away and, startled, it took off in typically slow cumbersome fashion. Later we cautiously approached the same area, hoping to surprise the eagle again, but this time the crows had got in first. There were at least a dozen around the carcass and the eagle was sitting in a tree presumably hoping to pick up the leftovers.

MAY 7

In the Narrow-leaved Peppermint forest on the Khancoban–Kiandra road I found a Beef Steak Fungus, which gets its name from its resemblance to a piece of beef steak (in France it is called an ox tongue). The colour and texture are much like meat, but the comparison ends there if you are game enough to sample it. Apparently it is tough and leathery and has an unpleasant flavour.

I also saw some cup fungi which looked rather like stale curled pieces of orange peel. These are Ascomycetes, so called because the spores (usually eight) are held in an elongated container, the ascus. The asci are borne on the surface of the cup. There were also some coral fungi, aptly named and with a range of colours from pure white through to deep purple. The spores are not borne on any specific organ but are found over the surface of the fruiting body. I have been told they are good eating in the early stages of growth and some are regarded as a delicacy in Europe.

I discovered the wings of a large moth (*Cheleptrix felderi*) lying on a path: grey leaf-like forewings and wavy patterned pink and brown hindwings. It had obviously been eaten by a Willy Wagtail, a bird of great aerial manoeuvrability and a

Beef Steak Fungus (*Fistulina hepatica*).

Cup fungus (*Peziza sp.*).

Cup fungus (*Chlorociboria aeruginascens*).

Toadstool (*Marasmius sp.*, formerly *Collybia elegans*).

Coral fungi: purple (*Ramaria sp.*) and white (*Clavulina sp.*).

A form of bracket fungus with gills (*Crepidotus sp.*).

notorious predator of insects. In my experience Willy Wagtails often return to the same spot, usually on a path or track, to eat their prey, leaving the wings behind. Sometimes this can be a useful means of gauging which species occur in an area. Thirty years ago I recorded the first evidence in Victoria of the Tailed Emperor butterfly from wings picked up on a track by a fellow walker.

I picked some blackberry leaves that looked as though they had been affected by rust. The orange pustules on the lower side of the leaves would be caused by the fungus Blackberry Orange Rust, a species accidentally introduced many years ago. Just recently we heard the startling news that a much more virulent rust, Blackberry Leaf Rust, had been found on blackberries in Victoria. This is the same species, but not the particular strain, which is being developed by Australian authorities working in Europe for eventual release in south-eastern Australia to control this invasive weed. Apparently the distribution of the infected blackberries indicates the rust has almost certainly been deliberately and illegally introduced. It is disturbing that this should happen, no matter how well intentioned the individual concerned. Before biological control agents are introduced they must be subjected to exhaustive research to show that they are not only effective on the target insect or weed but will not harm local plants and animals.

Later this afternoon I took the children up to Round Mountain and we camped overnight at the 1500-metre level on a ridge below the summit. We made our run rather late but managed to get a fire going and the tent up just as darkness fell.

MAY 8

Awoke to a frosty scene this morning with everything white, and the temperature had to be several degrees below zero.

We walked along the track which curls around the eastern flank of Round Mountain. Where loose earth covered the track and just beneath the surface masses of elongated ice crystals reared up like so many minute spires. It was a good example of how frost churns up bare ground. Wherever this 'needle ice' appeared, usually over areas of 2 or 3 square metres, the ground looked as though it had been roughly raked over.

A few grasshoppers were climbing sleepily up grass stems to catch the early rays of the sun. At this elevation insects have to adapt to a wide range of temperature.

Frost heaving—where needle ice pushes up and breaks the soil surface.

Leaf of snow gum (*Eucalyptus pauciflora*) covered with ice crystals.

Being small they have a high surface to volume ratio and so gain or lose heat readily. In full sunlight the temperature on the surface of a stone may reach over 30°C, even in autumn, but during the night may fall below zero. Alpine grasshoppers in some instances have assumed a dark or melanic colour, enabling them to absorb heat from the sun's rays rapidly during the morning, thus becoming mobile before other insects and gaining a competitive advantage.

One local species of grasshopper has the extraordinary ability to change its colour to accommodate the widely fluctuating temperatures. This is achieved by the migration of pigment granules in the epidermal layers towards the inner or outer surface depending on temperature. The alpine grasshopper (*Kosciuscola tristis*), found most commonly above 1800 metres, is blue-green above 25°C and black below 15°C. Early in the morning, towards evening, or in cold periods during the day its dark colour phase gives it the advantage of a higher internal temperature. On the other hand, in sunlight during the warmer parts of the day in its paler uniform it is less prone to overheating than black grasshoppers that cannot change colour, has more flexibility to move about and less need to seek shade.

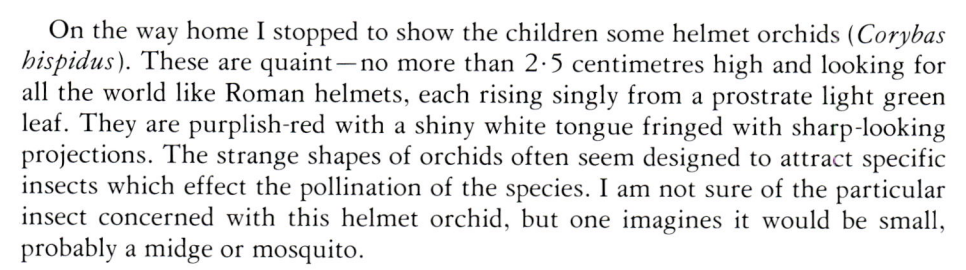

Dark coloured alpine grasshoppers (*Monistria sp.*).

On the way home I stopped to show the children some helmet orchids (*Corybas hispidus*). These are quaint—no more than 2·5 centimetres high and looking for all the world like Roman helmets, each rising singly from a prostrate light green leaf. They are purplish-red with a shiny white tongue fringed with sharp-looking projections. The strange shapes of orchids often seem designed to attract specific insects which effect the pollination of the species. I am not sure of the particular insect concerned with this helmet orchid, but one imagines it would be small, probably a midge or mosquito.

Helmet orchid (*Corybas hispidus*).

MAY 9

Today I spent some time at the top end of the Greg Greg Fire Trail and there I encountered a companionable little Yellow Robin and his mate. Yellow Robins will often come close, particularly if you are digging, as they are insectivorous and always looking for the possibility of getting worms or beetles.

A few native plants are still in flower. Urn Heath is quite common, with its strange pitcher-shaped flowers mounted in clusters on stiff prickly stems. I also saw a few individual plants of coral heath in bloom, an occasional yellow guineaflower, native violets and a straggling form of boronia, easily identified by its white four-petalled flowers.

Close-up detail of Urn Heath (*Melichrus urceolatus*).

Willows near the Murray.

May 10

Went out today with the purpose of photographing lichens. I have located a particularly rich display on a small granite outcrop, close by a stand of Alpine Ash.

The lichen is a composite organism, predominantly made up of a fungus, and of an alga. The vegetative part is known as the thallus and there are a great variety of forms, classed broadly as crustose, fruticose and foliose. The alga contributes sugars made by photosynthesis and the fungus provides some physical protection, nutrients and moisture. It is often quoted as an example of symbiosis, but the advantages enjoyed seem to be much in favour of the fungal partner. The fungus extends the range of habitats in which the alga can survive, but other descriptions of the relationship have included 'beneficial slavery' and 'controlled parasitism'.

Lichens have a great capacity to withstand the extremes of temperature on open rock surfaces. During the summer here all the exposed boulders are covered with what appear to be dead lichens, but once the rain and cooler weather of autumn return they begin to flourish again.

The whole existence of lichens when compared to higher plants seems most precarious. They have no capacity to take up water from below the ground and are reliant on moisture in the air. In spite of this, some can thrive perfectly well in the desert, making use of the dew they receive in the morning. In Chile one species exists comfortably in an area which has received no rain for hundreds of years, apparently absorbing sufficient moisture from the atmosphere. As well as moisture and nutrients lichens also absorb contaminants from the air and in cities will be readily killed by pollutants such as sulphur dioxide.

In rocky places lichens are important contributors to the formation of soil. The acids produced by their fungal component and the physical burrowing of the fungal hyphae, contribute to the natural slow disintegration of rock surfaces. Eventually a piece of rock splits away and another lichen will invade the newly exposed bare surface. Mosses are often successors to lichens as soil begins to form and in turn are followed by small broad-leaved plants and grasses.

I took some time to walk through the adjoining small stand of Alpine Ash. This tree is often called Woollybutt because of the dense fibrous bark on the lower part of the trunk, which runs out at 5 to 10 metres, leaving a clean light-coloured bole. With trees of perhaps 35 metres in height this was a modest stand (mature trees in ideal sites may reach 70 metres). It had clearly been subjected to relatively cool ground fires which had left the trees largely unaffected but had stimulated a dense undergrowth of Hop Bitter-pea.

Alpine Ash is a major source of high quality hardwood so most of the stands have been logged and are now in an immature stage following regeneration.

The original uniform stands of mature Alpine Ash are believed to have resulted from severe wildfires 100 to 200 years ago. For although this thin-barked tree is fire sensitive its seeds will not readily germinate, and have a chance of surviving, unless a suitable seed bed with sufficient light is available—which a hot fire provides. Unlike many eucalypt species, Alpine Ash does not regenerate by lignotubers, nor will it coppice readily. Regeneration occurs from seeds stored in capsules in the canopy, often of several successive years' flowering, ensuring an adequate supply from a species which produces widely varying amounts from year to year. Between 1 and 2 million seeds per hectare are dropped each season, though most are taken by insects.

In this species a fine balance seems to exist between its survival or elimination from a site. In nature the occasional devastating wildfire has been essential for the Alpine Ash. Little regeneration is possible without a major fire, yet the tree itself has no capacity to withstand it. But the seed held in capsules below the foliage and subjected to a short sharp burst of intense heat survives and later falls into an ash seed bed which provides the physical shelter, nutrients and micro-flora ideal for its germination. However, should another major wildfire devastate the new stand before it has reached seed-bearing age at about ten years, the species will be eliminated from the area.

Foresters aim to simulate nature on a small scale, clear felling limited areas after logging, burning the residues and sowing seed by hand or air, thus ensuring an even-age stand of high quality timber.

A foliose type of lichen, *Parmelia rutidota*.

A fruticose type of lichen, *Cladonia chlorophaea*.

A crustose form, the brown lichen, *Protoparmelia petraeoides*.

May 26

I awoke this morning to find visibility down to a few metres. It was extremely foggy during the drive up from Melbourne last night and I had to travel very slowly over the last 30 kilometres. This is real winter weather although officially the season has not arrived. The fogs are dense, heavy and there is an all-pervading feeling of dampness. They sometimes last for the best part of the day. Today was no exception, and the sun didn't get through until 1 o'clock.

Dense morning fog.

It has been a strange year so far. Following the 1982–83 drought we had good rains and no really dry spells, and this continued until early April. But now the weather has turned dry again and is causing concern. We've had only 8 millimetres of rain in the last six weeks whereas normally we would expect around 75 millimetres. Because of the good early autumn rains the subsoil is still moist, but newly sown oat crops and pastures are feeling the 'pinch' and although the paddocks all look green and healthy it is now too cold to expect any real pasture growth until next September. By July we will need to feed out hay heavily to the cows and calves.

There was little I could do in the fog so I pottered about near the cottage. Around the Apple Box trees I found many nettles, with caterpillars of the Admiral butterfly in curled leaves. Nettles are their food plant and they live over winter as caterpillars, each one bending over a leaf to form a small cylindrical shelter. With the added protection of stinging hairs on the leaf, the caterpillar appears relatively safe from parasites and predators.

I walked above the cottage to the rocky slopes of the Lighthouse Mountain where there are a number of small cypress-like plants. These are Wild Cherries, sometimes known as Cherry Ballart. The plant produces a fruit but the edible part is really the bright red fleshy stalk which supports the smaller green fruit. They are tiny and you would need to pick a great number to make a worthwhile meal, though they are said to have been one of the foods of the Aborigines. The Wild Cherry is a partial parasite, with many roots connected to those of surrounding eucalypts and wattles.

In a gully on the Kiandra road, near the Narrow-leaved Peppermint forest, there were also a number of fungi of the genus *Cortinarius*. They are typical toadstools but have a characteristic veil (or cortina) which covers the gills during the development of the cap. The remnants of this veil adhere to the stem when the mushroom assumes its mature shape.

It is the range of colour—from the richest and deepest reds, through purples, orange, yellow, and even an unusual colour for toadstools, green—that is the remarkable and attractive feature of the cortinars. They also exhibit a great variety of texture, sometimes covered with liquid and a highly enamelled sheen. You see the glossy caps emerging through the leaves and twigs on the forest floor with small pieces of stick, grass and soil attached to them.

Many of the cortinars and other showy agarics, such as *Russula* and *Amanita*, are biologically associated with the trees under which they grow. So also are a great variety of less spectacular forms and some that rarely produce a fruiting body above the ground. The association is achieved by the fungal roots, or mycorrhizae, which form a mantle or sheath around the tiny lateral roots of the tree and penetrate between the cortical cells. Through their mycorrhizae the fungi assist the tree in obtaining essential nutrients from the soil, especially phosphorus, an element so often in short supply in Australian soils. They are also believed to aid in the uptake of water, the preservation of minerals against loss through leaching, and in reducing the potentially lethal effects of salt and other toxic substances.

Many species of fungi associate with a wide range of trees but some have more specific relationships. One of the best known examples of an obligate affiliation is the brilliant red and white dotted Fly Agaric, a native of the northern hemisphere. It exhibits a close affinity to pines, though it is also less commonly found on silver birches and chestnuts.

Not all fungi live such a cooperative existence with trees and one I saw today, a mass of yellow toadstools (*Armillaria luteobubalina*) at the base of a tree stump, is a serious parasite of eucalypts. It spreads by long thread-like rhizomorphs, penetrates the roots of the tree, moves up inside the bark and kills the vital cambium tissue. The tree stumps of previously infected trees play an important role in the ecology of this fungus as they are a source of infection for healthy trees.

Towards sunset I stopped at Bradneys Gap to look at a stand of blue gums (Eurabbie) and some Mountain Swamp Gums. The immature foliage of the blue gum, being blue and with a square-sectioned stem, differs markedly from the mature foliage. The large knobbly buds also have a bluish waxy appearance. The tree grows to a great height and is useful in providing bulky chunks of timber for building. It also has an extraordinarily fast rate of growth in the immature stage. It maintains maximum photosynthetic activity, and at the same time transpires vigorously, drawing heavily on the soil's water reserves as many gardeners have discovered to their cost. If you grow a blue gum near a house on a soil type which shrinks as water is withdrawn, you may find cracks appearing in the walls as the foundations move.

On the positive side, blue gums have proved very useful in draining swamps in India and other countries where malaria is a problem.

I noticed many young red gums springing up along the side of the road. They were *Eucalyptus blakelyi*, which have broad oval immature leaves. (The River Red Gum, *E. camaldulensis*, is easily distinguished at this stage because it has narrow lanceolate immature foliage.) These small seedling trees are thriving because they are not exposed to any grazing pressure. In the paddocks on the other side of the fence there were no signs of seedlings. Later as I drove home I followed up this observation and was pleased to see that there has been considerable germination along a number of roads. This demonstrates the quick response Australian native plants make to favourable conditions. This last summer and autumn has been the best for years and the hardy seedlings of wattles and eucalypts are flourishing.

Green toadstools (*Cortinarius austrovenetus*).

Armillaria luteobubalina, a parasite in eucalypt forests.

Pixies Parasol (*Mycena interrupta*).

MAY 27

I woke again to heavy fog. I browsed around the garden and heard a noisy pair of Yellow-tufted Honeyeaters. If they are looking for flowering plants they won't find too many here, although insects are said to be their principal food.

Later in the morning, as the fog began to lift, I could vaguely see a flock of birds flying around the dam. They proved to be Welcome Swallows, displaying their forked tails and erratic yet graceful flight patterns. At first glance one might think they were looking at their own reflections, but swallows have a particular interest in insects associated with water—mayflies, gnats and mosquitoes—and are noted for their skill in catching them on the wing. I watched for half an hour or more and never once did they land, wheeling in tight circles, occasionally touching the surface of the water with their feet or with one wing and creating little ripples.

After lunch the sun finally broke through the fog and I set out along the edge of Welumba Creek, making for the lower end of the Greg Greg Fire Trail. On the way I noticed a flock of Eastern Rosellas. They are beautiful birds, apple green, red and yellow, with flashes of blue. They livened up the sombre, misty scene as they fed on the seed heads of Saffron Thistles, which also seem attractive to cockatoos.

The Welumba Creek area, where the Greg Greg Fire Trail begins its abrupt ascent towards the snow-line, is a mosaic of rocky outcrops dissected by deep gullies with damp shrubby patches on the southern aspects. Dramatic changes in vegetation take place in the space of a few metres.

One of the forested areas was thickly carpeted with native grasses. Here, in a green dell among the trees, I discovered a curious fungus, commonly known as a 'vegetable caterpillar'. This particular species, found at the base of wattles, was *Cordyceps gunnii*, yellow-green in colour with a dark, almost black tip. The whole of the fruiting body stands about 7 centimetres above the ground and is long and club-shaped. I dug it up and 5 centimetres below the fruiting body it was attached to a dead swift moth caterpillar. A year or so ago the fungus would have invaded the caterpillar, its fungal threads proliferating and digesting the internal tissue of the caterpillar before the fruiting body emerged from the soil.

In mediaeval times, when witchcraft was rampant, vegetable caterpillars were seen as a manifestation of the belief that animal matter could change spontaneously to plant matter. Here was a caterpillar from which sprouted a growing plant. It was not until late in the 18th century that biologists were able to explain that fungi invaded the caterpillar while it was alive and ultimately produced a spore-forming body above the ground. In most species the fruiting body is formed on the larva or pupa, but in some it appears on the adult insect, although to my knowledge this has not been observed in Australia.

An interesting feature of the fungus is that certain species select particular larvae of moths, beetles, or wasps to parasitise. In the tropics cordyceps have been found growing from the head of adult wasps. In the 18th century these were something of a collector's item and were known as vegetable flies.

I quietly crept towards a bird calling further up the hill. Although I did not see it, I am sure it was a lyrebird. It was not producing the range of mimicry so often heard from the male bird, but was making the characteristic 'choo' sound which is its territorial call. I had not expected to find lyrebirds here as it seemed too dry and stony. However, on closer examination I found disturbed ground where grasses, herbs and mosses had been worked over extensively—typical of lyrebird scratchings as it searches for small crustaceans, insects and worms.

Later in the afternoon I encountered the communication systems that exist between kangaroos. As soon as the first kangaroo saw me it seemed noiselessly to transmit its concern to its neighbours, although the alarm system is said to be a cough. Their disquiet was relayed to the cattle as well. The stock feeding near by quickly became aware that the kangaroos were looking at something and also turned their heads, seeking the source of agitation. At another stage, as I traversed a small valley, I heard some heavy thumping on the ground from the tail of a large old man kangaroo about 100 metres away. This quickly alerted all the others for a

Vegetable caterpillar (*Cordyceps gunnii*).

Galerina hypnorum growing among moss.

Toadstool (unidentified).

long way around and prevented me from getting too close. Finally I did get a picture of a group immediately below me in a grassy gully. They were unaware of my presence, while those further afield could see me but seemed unable to communicate with them.

Coming home in the semi-dark I walked through an area of the introduced pasture plant phalaris, in a place fenced out from cattle and sheep. The seed heads stood on dry stalks over a metre high. Phalaris, introduced to Australia from the Mediterranean region, is a valuable pasture grass which is sown widely in south-eastern Australia. It is deep rooting and able to tap sources of underground moisture even when the surface is very dry.

Sunset reflected in a dam below Cochrans Gap.

In a prolonged hot summer here the leaves dry up but the plant survives by means of subterranean dormant buds which obtain water from the soil through the deeply penetrating roots. In my view phalaris is clearly the best pasture grass for the rising country of quick-drying granitic soils. It survives where more nutritious species such as Perennial Rye Grass die out, and provides vital winter feed and protection against soil erosion.

Phalaris has been the subject of much research by the CSIRO, which has produced more vigorous and drought resistant strains. It has also developed varieties with a lower alkaloid content, because one of the problems with phalaris is that at certain times of the year, particularly in the autumn following the first rains, its content of these chemicals can be sufficiently high to poison stock.

With all its great capacity for growth and adaptability to our environment, phalaris has one surprising weakness. It produces millions of tiny seeds per hectare yet few survive in the competitive environment of the paddock. Once established there one can almost guarantee that the plants will never increase in number. Paradoxically, it may proliferate along the side of roads where the fertility level is sufficient for the phalaris but does not encourage competitors.

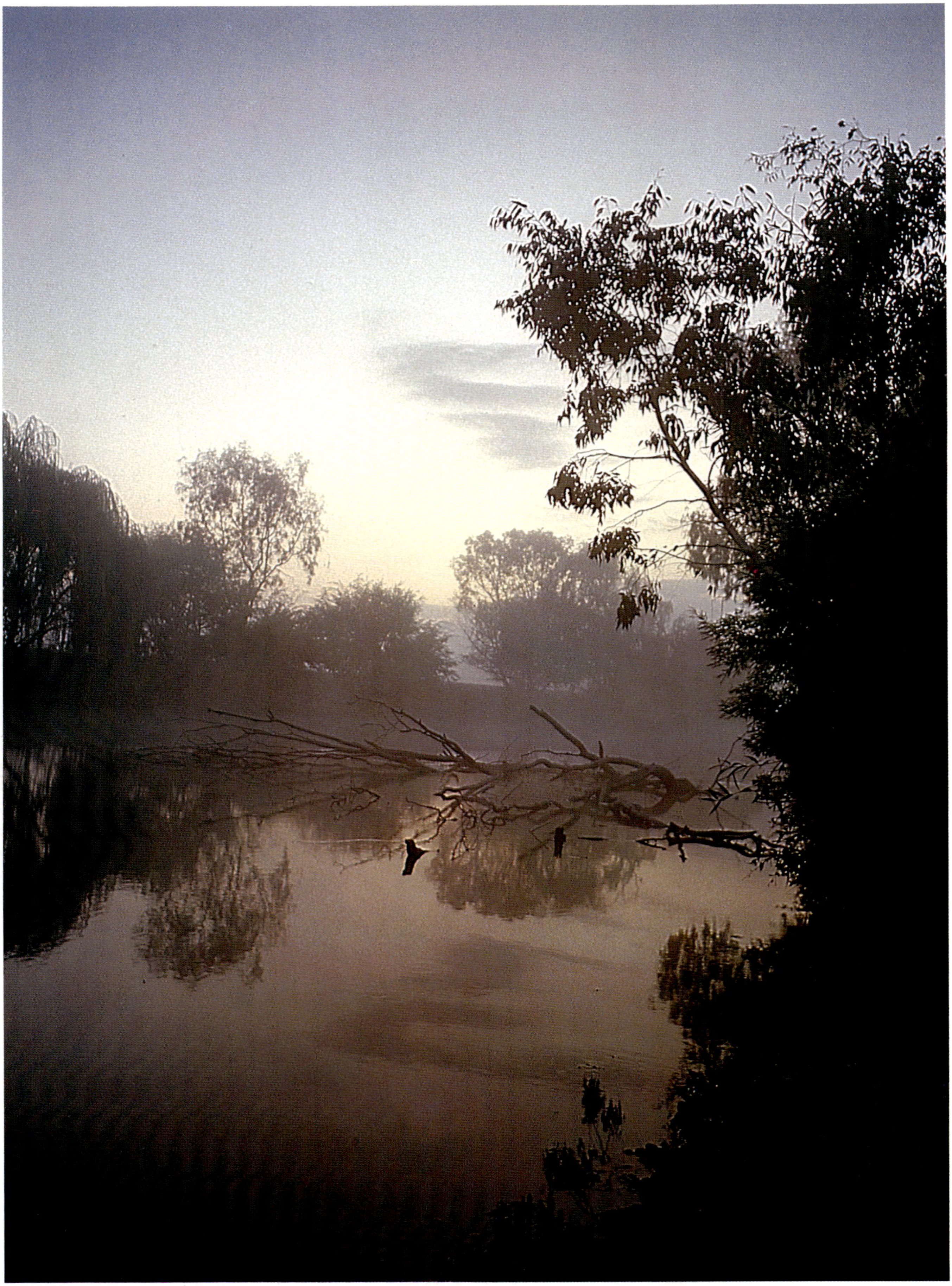

May 28

The fog, if anything, was even thicker today. I decided to visit the Tooma River, which is about 100 metres lower than the cottage, and here the visibility was relatively good beneath the fog. It would seem these heavy winter fogs sometimes rise about 50 metres above the Murray and Tooma and hang as a band—perhaps 150 to 300 metres thick—for a period of several hours before the sun's temperature finally disperses them.

The autumn colours of the deciduous trees are not as vivid this year which may well be due to the run of heavy fogs. Bursts of warm sunny weather seem to provide the best conditions. Nonetheless, the meandering course of the Tooma is now neatly outlined in golden willows and the gardens around the homesteads, with their poplars, Liquidambars, ashes and oaks, provide a kaleidoscope of colour.

I continued on to Welumba Creek, to a point where the stream drops suddenly before running out into the gently sloping farmland. You can hear the roar of the waterfall from the Greg Greg Fire Trail, but it is only after descending about 100 metres that you can appreciate the rocky and precipitous character of the ravine. On the northern aspect, facing the sun, the vegetation is typical of that on a stony skeletal soil, with several species of wattle and heath-like plants. On the other side of the gorge, merely a few metres away but facing the south, you can find a quite different environment—pockets of pomaderris, ferns and moss. The sun does not reach these small sheltered stretches of the creek at all during winter.

Autumn colours, Possum Point, Tooma River.

I found a small rock cave with a number of mason wasp nests on its walls. I picked one off and put it in my pocket. Later, when I opened it up, I found a number of interesting items. Mason wasps (Eumenidae) are solitary, and parasitic, mainly on the caterpillars of moths. The female wasp skillfully paralyses the caterpillar with its sting and takes it back to the nest. It makes individual chambers in the clay structure, in which the caterpillars are placed, and these comatose caterpillars provide food for the wasp's own larvae when they emerge.

Morning fog lifting over the west face of the Lighthouse Mountain.

OPPOSITE PAGE:
Dawn, Tooma River.

When I took the nest to pieces there were four wasp larvae, three active spiders that had found a home there, and some *Anthrenus*. The latter are larvae of museum beetles, which clean up the organic matter remaining after the wasp's larvae have eaten the caterpillars. In addition there were a number of tiny cocoons of the braconid wasp, also parasites of caterpillars of moths and butterflies, and I can only assume that one of the caterpillars brought in by the female mason wasp had previously been parasitised by braconid wasps. These must have emerged within the nest and found their way out, since all the cocoons were empty and there were no signs of adult wasps.

View over Taylors Creek towards Welumba Hill and the China Walls.

MAY 29

Morning dew on spider web.

The usual foggy outlook. On the verandah dew drops had formed on the tendrils of the vines and on the spider webs. In fact, wherever I looked there were tiny webs. Each barb on the barbed wire fence had its strands of gossamer and dew. You tend only to notice these webs early in the morning because of their glistening cover of dew drops. Later in the day you don't see them, although they are always particularly thick on old thistle heads. The thing that surprises me is that I can never find the spiders themselves—and there must be thousands—but I guess they are very small and presumably nocturnal.

In the afternoon when the fog rose I went back to look for the lyrebird I had heard two days ago. Although I didn't actually see the bird, I did hear much of its repertoire of mimicked bird calls, which included cockatoos (both the Sulphur-crested and the Gang Gang), Crimson Rosellas, kookaburras and a variety of other birds, all in quick succession.

JUNE 19

A small flock of Sulphur-crested Cockatoos above the Tooma River.

Red-necked Wallaby.

The weather continues extremely dry. We've had barely more than 25 millimetres of rain in eight weeks and would normally expect 150 millimetres. It was very fortunate that we had such good early autumn rain.

I walked through Narrow-leaved Peppermint forest on Pierces Fire Trail, which leads back towards the farm and passes over an old forest grazing area called Spring Tops. I didn't get that far but came across an open plain of around 3 hectares covered with Kangaroo Grass. The area appears to have been cleared at some stage although no one seems certain whether the opening is natural or man-made. Around the edge of the plain and very shy was a group of Red-necked Wallabies. Such small clearings favour wallabies for as long as there is sufficient cover in the form of nearby forest or undergrowth, the extra grass afforded by the clearing enables them to breed up. Red-necked Wallabies are quite distinct from Eastern Grey Kangaroos, but the differences are not as radical as one might expect and primarily are those of size, and also colour and conformation. Presumably these differences have evolved because of their different habitats. Wallabies don't seek the open country unlike the kangaroos and tend to go under objects rather than over them. Thus they are shorter, dumpier, and less quick and athletic in their movements. In other ways, such as the mode of birth and the rearing of their young in and out of the pouch, they are similar to kangaroos.

JUNE 20

I spent time at the cattle yards this morning where calves were being marked. These are our autumn calves—usually the best we have, reflecting the seasonal pattern in this area and the way the pastures grow. Autumn-dropped calves are at a sufficiently advanced stage of development in spring to take full advantage of the lush growth of grass and clover. In most seasons they sell better than the later-dropped spring calves. However, with a preponderance of autumn calves you are more vulnerable to a late season, when the autumn rains fall at a time when the temperatures and light are insufficient for adequate pasture growth to see the cows and calves through a hard winter, and you have to feed out much more hay.

Young cattle on closely-cropped winter pasture.

I picked up a eucalypt leaf adorned with several shiny red spherical growths, just like small Jonathan apples. These were plant galls and when I opened one up there was a tiny scale insect inside.

Galls are abnormal growths produced by the host plant in response to the presence of an invading organism—often a species of insect, but sometimes a bacteria or fungus. The advantages of the association seem to be for the invader, which has a protective shelter specially built for it, but the plant may also gain from the gall by 'walling off' the 'enemy' and localising the damage.

A great variety of insects are responsible for the production of galls, including wasps, flies, thrips, scale insects, beetles and moths. In Australia the coccids or scale insects are responsible for some of the largest and most bizarre forms. Galls may be produced on the leaves, stems or buds and their shapes may resemble grapes, pears, leaves and even nutmegs, each being specific to a particular gall-forming insect and the plant's response to it.

The male scale insect is small, with tightly folded wings, in contrast to the much larger, wingless and soft-bodied female who resides permanently inside her gall. The male enters the female's gall by a tunnel to mate.

Galls on the leaves, stems and flower buds of eucalypts are often formed by small flies (*Fergusonina*) in association with nematodes. The fly establishes the infestation, the nematodes emerge from the fly larvae and together they contribute to the formation of the gall. In due course, after the female nematodes have been fertilised, they enter the fly larvae and produce larval nematodes which invade the developing ovaries. After the fly has emerged from the gall the larval nematodes are returned to the plant through the oviducts as the female fly lays her eggs on the plant — and so the cycle is repeated.

A gall of the bud of an Apple Box (*Eucalyptus bridgesiana*).

The same gall dissected to show the larvae (Diptera) within.

Insect galls on eucalypt leaf.

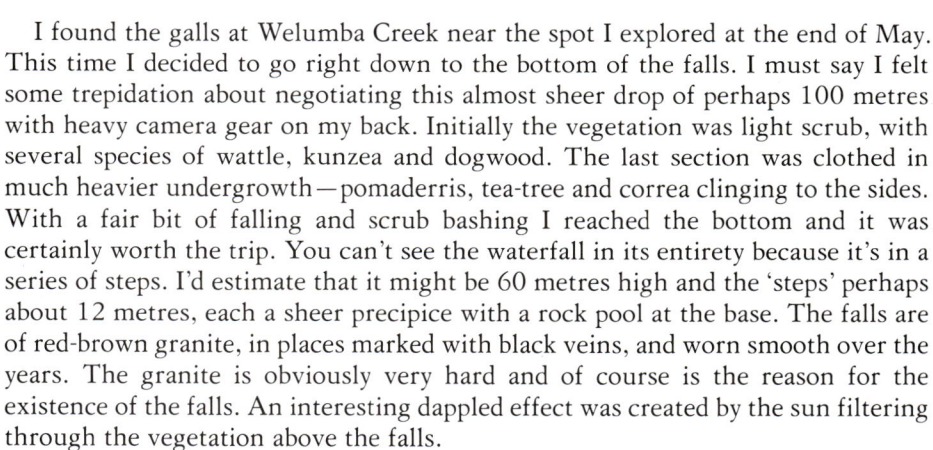

I found the galls at Welumba Creek near the spot I explored at the end of May. This time I decided to go right down to the bottom of the falls. I must say I felt some trepidation about negotiating this almost sheer drop of perhaps 100 metres with heavy camera gear on my back. Initially the vegetation was light scrub, with several species of wattle, kunzea and dogwood. The last section was clothed in much heavier undergrowth — pomaderris, tea-tree and correa clinging to the sides. With a fair bit of falling and scrub bashing I reached the bottom and it was certainly worth the trip. You can't see the waterfall in its entirety because it's in a series of steps. I'd estimate that it might be 60 metres high and the 'steps' perhaps about 12 metres, each a sheer precipice with a rock pool at the base. The falls are of red-brown granite, in places marked with black veins, and worn smooth over the years. The granite is obviously very hard and of course is the reason for the existence of the falls. An interesting dappled effect was created by the sun filtering through the vegetation above the falls.

Behind me some blackberries were still fruiting, though I doubt whether the rose-red fruit would ripen now. I guess that down here, with so little light, they have been frustrated and have developed much more slowly than their sun-spoiled colleagues further up the hill.

Later on, after attempting some shots of the Dargals in the setting sun, I picked my way back through the paddocks, creeks and fallen timber in the dark. A fox with a large bushy tail crossed my path, obviously well fed, probably on rabbits. We see a lot of foxes here but don't regard them as a problem. In years gone by they were held in suspicion by farmers and to a lesser degree this is probably still the case. Research indicates that they are essentially carrion feeders rather than killers of lambs. Perhaps more surprising is that insects make up a significant proportion of their diet and they also eat blackberries and other fruit. There is no doubt, however, that they are killers of chickens if they can get into a poultry enclosure.

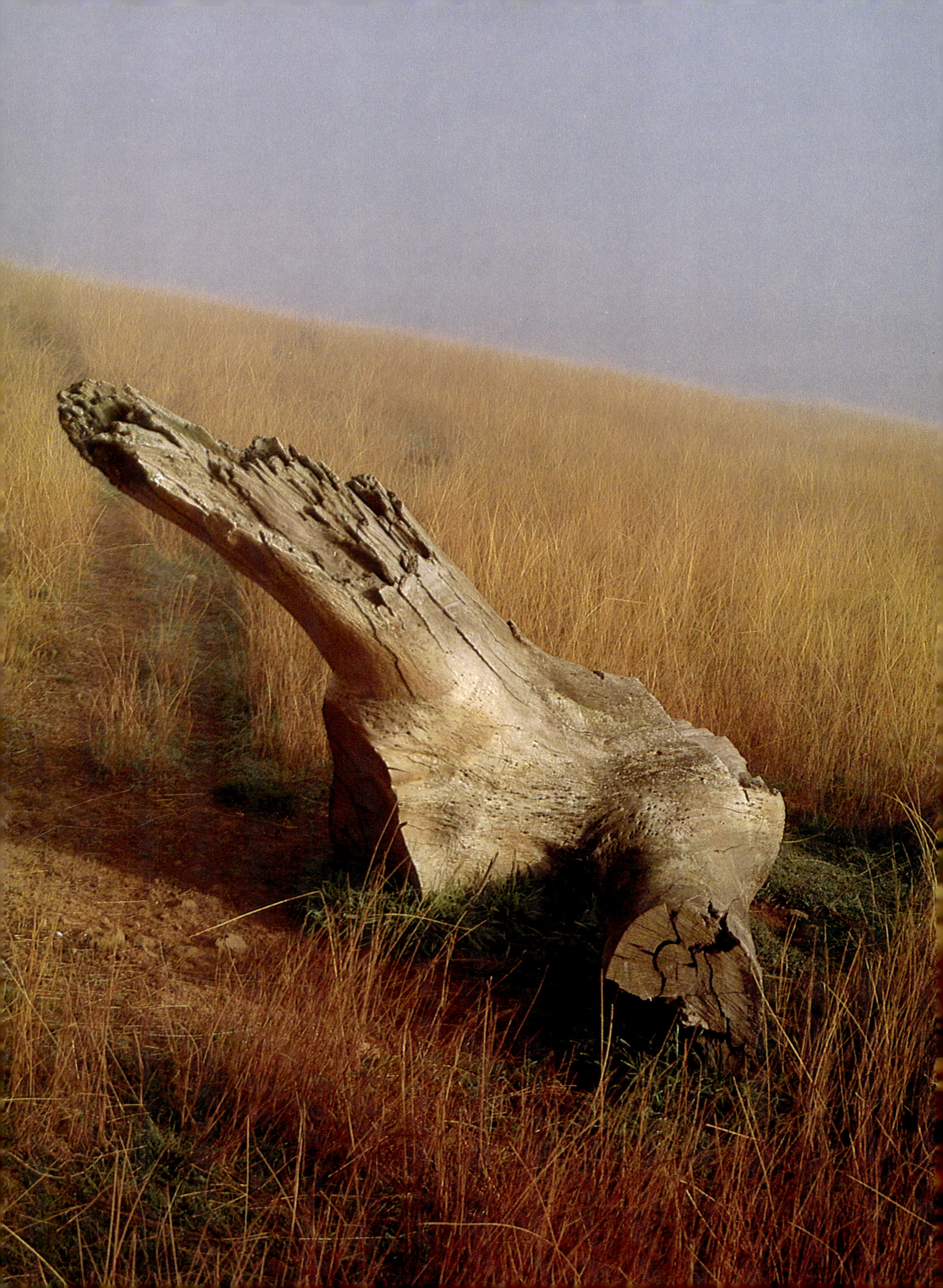

June 21

I left at first light for Mount Elliott, famous as a gold-field late last century and located just south of the Murray River near the township of Towong. I trudged up the track and hoped that I could get above the low-lying cloud to catch the first rays of the rising sun.

I had to make my way through a sheep camp at one stage. I wondered why sheep always choose a rising piece of ground, usually the top of a hill, to camp. I guess that it gives them a defensive advantage over a predator and the ground is better drained and therefore likely to be warmer.

Sheep camps are a nuisance on farms because they result in a local increase in soil fertility which in turn encourages weeds. In the spring months you can fly over parts of the eastern highlands of Australia and pick out sheep camps by the yellow patches of Capeweed, almost invariably on ridges or the tops of hills.

Suddenly I was out of the fog with the Dargals and Grey Mare Range just visible across the valley, between two bands of cloud. It was a quite unworldly scene with the upper layer diffusing light from the rising sun, converting it to a rich golden haze and illuminating the surface of the lower cloud.

The sun illuminating cloud layers, looking east from Mount Elliott.

This afternoon I walked to Cooaninnie from the transmission line and found it fairly easy to reach the first patch of cleared country. Cooaninnie now comprises two small areas of native pasture in an otherwise continuous forest which stretches up to the Dargals. It was cleared last century by the Harrison brothers, Jack and George, prior to the family selecting property near Khancoban, where their descendants still live.

Cooaninnie was also later a base camp for the Toolong Gold Field, located on a narrow lead on Dargal Creek, a tributary of the Tooma River. Situated at an elevation of 1300 metres immediately east of Big Dargal it must have been a particularly bitter place during winter.

OPPOSITE PAGE:
Remains of tree trunk among red grass (*Bothriochloa macra*) on the slopes of Mount Elliott.

The rush began in 1893, six years after the discovery of gold, and in that year it is recorded that part of the surface lead the men were working was covered with 2 metres of snow. As winter set in most of the miners retreated to the more equable climate below, but some thirty braved it out in rough huts. At its peak in 1894 three hundred miners were located at Toolong on Dargal Creek and several small nuggets, including one of 11 ounces, were found. But it was a very limited find and within two years most of the miners had left, although small groups of fossickers searched for gold until 1902. Cooaninnie, which is located on the old stock route up to the snow leases, continued to be used until twenty years ago to muster cattle and as a stop-over point before the tough climb up the spur to the snowgrass country.

The second and larger area of cleared country at Cooaninnie was more difficult to reach, but I got there after about an hour's walking. The country here is quite open, with signs of past grazing, but from my fairly cursory look is clear of weeds. Nor does there seem to be much reinvasion of the cleared flats by natural vegetation. One would imagine that the Silver Wattles would recolonise fairly quickly but it is probably not shaded enough for them. The area seems to have stabilised on a Kangaroo Grass-based pasture, which the kangaroos and rabbits keep well trimmed now. When you unexpectedly emerge from the forest on to these clearings, the image is one of a tranquil rural scene, and you expect to see farm buildings and sheep and cattle.

I walked through some beautiful open forest country. It was made up of Narrow-leaved Peppermint interspersed with white-trunked Candlebarks, and an undergrowth of Silver Banksia, bush peas, Handsome Flat-pea, Coral-pea, guinea-flower and tussock grass. It will be very colourful in spring.

One Coral-pea (or Sarsaparilla) was already coming into flower. Does this indicate an early spring? Normally Coral-pea does not flower here until September.

The kangaroos seemed more curious than frightened when I arrived on the flats. They probably had not seen many humans before. I decided about 1 o'clock to make my way back to the road. Being a confident bush walker I had not bothered to bring a compass or map on this short trip and that proved a mistake. I crossed my tracks at some stage and got off course, and it took me about three and a half hours to get to the road. On the way I sprained my ankle while clambering over a slippery log in a fern gully.

On the trip back I noted that the wombats had been busy. There were dozens of burrows in the light gravelly red soil and usually among thick bracken, and it was difficult to get through without falling into the openings.

At one point I took a rest and turned over some stones. Two orange cockroaches slowly emerged, obviously feeling the effect of the cold weather. Most cockroaches are nocturnal and dark brown, or even black, but there are a number of species that move about during the day, some in glinting shades of metallic green, but in my experience they are so fast-moving that you rarely get the chance to observe them. I recall seeing a large one last summer, resplendent in black and green, sunning itself on a bossiaea bush. My view of it lasted less than a second before it vanished.

I came across a few lyrebird dancing mounds amid the bracken, obviously recently used. I wondered how they performed in such dense undergrowth and how they could hope to attract a female audience. The ladies would have to have seats in the surrounding trees to see much of the action!

Coral-pea (*Hardenbergia violacea*).

JUNE 22

Today I decided to get out of the heavy morning fog as quickly as I could and so made my way up Mocattas Ridge, which rises steeply above Taylors Creek to a height of over 600 metres.

I was feeling the effects of my sprained ankle and after a slow and laborious climb, virtually on one leg, I stayed on the ridge for most of the day. Even at this

Above the cloud layer.

elevation the fog did not rise until after 1 o'clock, but down below on the Tooma River it didn't lift all day. From my vantage point above, the thick white fog in the winter sunshine stretched away on all sides to the horizon, broken only by occasional peaks, each looking like a small forested island.

Not being able to move around much and with little else to do, I was able to observe the immediate forest around the ridge. Here the bush is recovering from a small fire which occurred late last summer. I noted how quickly the plants had regenerated. Grasses and herbs are coming up, but the most noticeable to recover are the grass-trees. Their trunks, burnt coal black and still with the charred remains of the original bark, now are sprouting with bright green spiky leaves.

Grass-tree (*Xanthorrhoea australis*).

On the edges of the forest country I noticed a lot of Drooping Mistletoe (*Amyema pendula*) and it is very evident that this common parasite of eucalypts is much thicker on the margins than in the forest itself. The reason is that the seed is mainly distributed by the Mistletoe-bird, which prefers the open edges of the forest. Although small and seldom noticed, the male bird is actually brilliantly coloured—dark blue with a red breast and some white markings. The sticky seeds of the mistletoe are eaten by the birds but by a specialised mechanism they by-pass the stomach, where they would be damaged, and pass rapidly through the intestinal canal. A proportion of the seeds is then excreted on to branches of the host tree.

Seeds of some species of mistletoe are highly specific to one host and while they will germinate on other tree species, they will not develop the vital connections to the host's tissue. Mistletoes are essentially parasites of the xylem or water-transporting tissues of the plant, drawing on the host's reserves of water and minerals. They have their own capacity to photosynthesise, with leaves like other plants.

The clearing of the forest has led to the proliferation of mistletoe. In some situations this is a cause for concern. Mistletoe reduces the rate of growth of the host tree though it does not often kill it. Occasionally the mistletoe dies rather

Flowers of Drooping Mistletoe (*Amyema pendula*).

The shape of the Drooping Mistletoe leaf (left) closely resembles that of its eucalypt host, though its colour is quite distinct.

OPPOSITE PAGE:
The big frost.

Frozen spider web in the fence netting.

than the tree. Today I noticed both situations: a few trees killed by mistletoe and mistletoe dying on what appeared to be a healthy tree. In the latter case the water requirement of the mistletoe during the drought may have been higher than that of the host plant.

Drooping Mistletoe is easy to see because its olive green colour and closely bunched foliage contrast with the foliage of the host trees. Examining the mistletoe leaves today and comparing them with those of its host Red Stringybark I was impressed by the similarity in size and shape, despite the obvious difference in colour. I have always been intrigued by the apparent imitation by a mistletoe of its host, a well-developed feature of this parasitic group of plants in Australia.

It is proving a puzzle to find a satisfactory explanation for this phenomenon. Some of the 'imitations' are remarkable and none more so than that of *Amyema linophyllum*, a parasite of Buloke. So accurately does this species echo the unusual long needle-like shape of the stems of Buloke (the leaves are reduced to tiny scales) that it is only with some practice that you can find the mistletoe.

Why should this be so? The leaf shapes cannot have 'converged' through natural selection because of a common climate, since many shapes and leaf sizes can function effectively and two quite dissimilar mistletoe leaf forms can be found in the same area, each an imitation of its particular host plant.

One theory suggests that mistletoes have evolved a camouflaged or cryptic form to conceal themselves from herbivorous animals such as possums. However, the existence of many species such as the one I was examining today, where the concealment is 'destroyed' by colour and density of foliage, weakens this hypothesis. Possums have also been shown to be disinterested in eating mistletoe leaves.

Another idea is that close resemblance to the host may aid in more efficient dispersal of the seeds by the Mistletoe-bird. A mistletoe which closely resembles the host may cause the bird to use the host tree as a long distance sign, thereby improving the likelihood that the seed is dispersed on to the correct host.

Birds clearly play an important role in the ecology of mistletoe, both in the dispersal of pollen and of seed, and the mistletoe would appear to adopt opposing strategies in relation to each function. The mistletoe flower is almost always conspicuous and brightly coloured and of a quite different shape from the leaf. Pollen and nectar are produced over a relatively short period of time and there is presumably an advantage in quickly attracting birds to the flowers. The seeds, on the other hand, mature over a long period, timing is less important and the concealment strategy ensures that the Mistletoe-bird must hunt widely over the host trees to find the mistletoe, thus getting a maximum spread of voided seeds.

A quite different theory has been proposed which suggests that the similarity in leaf appearance is the result of evolution as the mistletoe has adjusted hormonally to become compatible with its host. For instance, the mistletoe has become adapted over time to its Buloke host, and the extraordinary lengthening and narrowing of the leaves of the mistletoe is a response to the same hormones influencing the elongation of the stem of the host, and has nothing to do with the influence of external environmental factors, such as predators (possums) or distribution of seed (Mistletoe-birds). Further research obviously is needed to clarify this issue.

JUNE 23

A bitterly cold morning −3°C in Corryong and the heaviest frost I have ever seen up here. It extended right up into the tall trees and gave the garden of the cottage a fairy-tale look. Behind the cottage on the Lighthouse Mountain tussocks of red grass left from last summer were encased in ice, like bundles of silver needles running up the hill. The wattles in the garden were sparkling with ice crystals and the fence netting was a lacework of frozen spider webs.

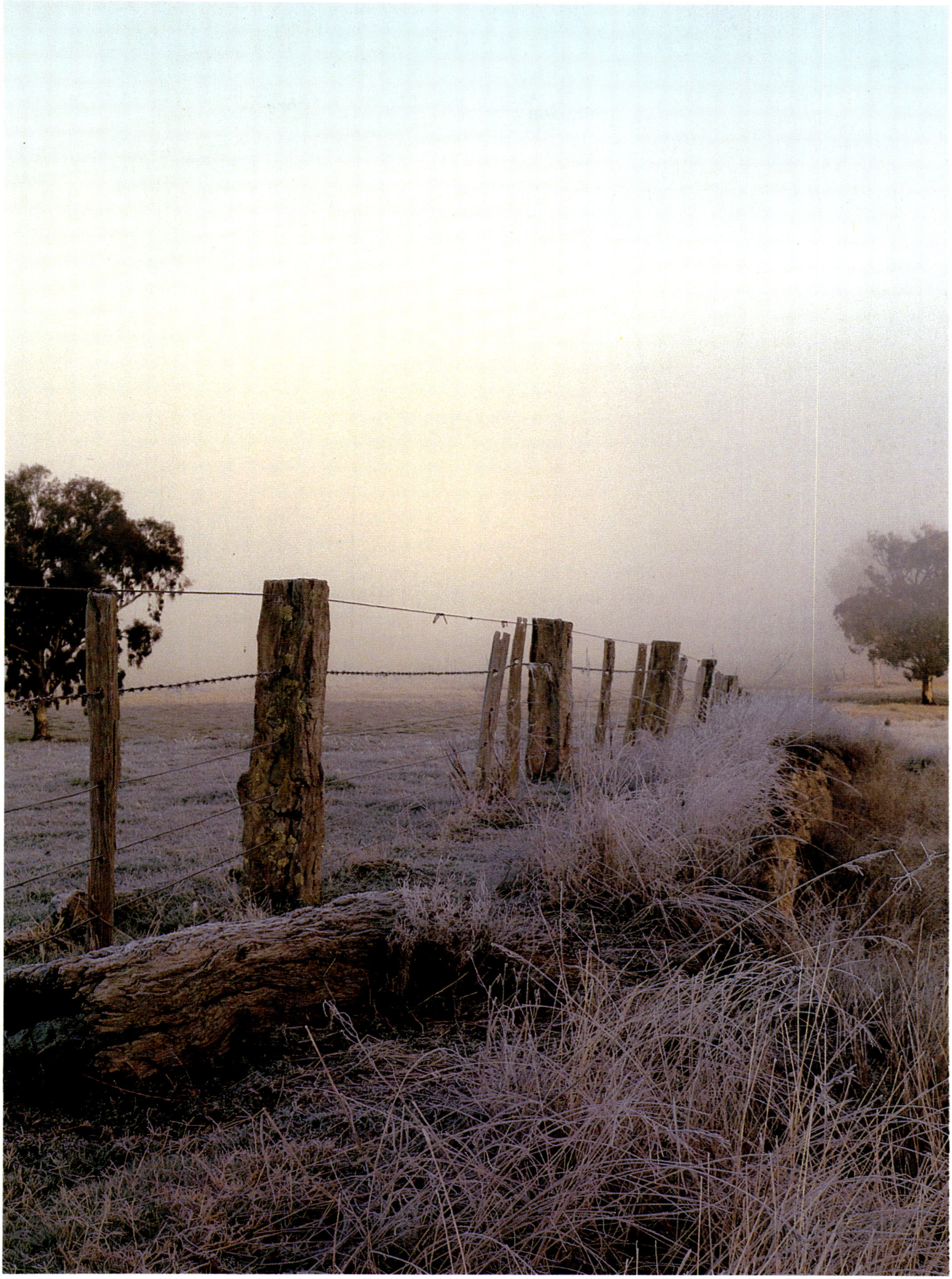

Cockchafer larva (Scarabaeoidea).

Flame Robin.

Eastern Swamphen, Towong.

White Egret by pond at Towong.

As I turned over the logs behind the cottage, hoping to find some bugs, I became aware that I was being observed by a number of male Flame Robins. They were clearly interested in the large and succulent cockchafer grubs that had been revealed. The robins became quite tame after a while and I was able to set up my camera on a tripod and take photographs of them. The Superb Blue Wrens, also insect feeders, did not show the same curiosity. They were mostly immature birds and probably lacked the confidence to come up as close as the robins.

Later on it warmed up a little and I went to the lagoons near Towong, in search of water birds. There were few — a dozen swans, several Eastern Swamphens and a White Egret. Unfortunately water birds on the river flats here have been subjected to a good deal of harassment and irresponsible people have undoubtedly shot some of the protected species. As a result the birds are very wary.

For many years, on each visit to the farm, we used to look forward to seeing a particular Brolga on the swamp near Towong Bridge. It became quite a tourist attraction before some 'sensitive soul' is said to have shot it, and so far no other Brolgas have taken its place.

JUNE 24

The fog was much lighter this morning and I climbed Mount Elliott once more.

On the way up I rolled over a rotting log and uncovered a large wolf spider, recognisable by its four eyes arranged in the form of a square and its long legs. They are fierce-looking spiders, though today this one was quite immobile in the freezing weather. The Australian species are said not to be venomous. We often find wolf spiders scampering round the house up here. They don't remain still on the wall like the huntsmen but are more likely to be seen running swiftly and rather erectly across the floor.

I have frequently found female wolf spiders with a mass of tiny spiderlings on their backs and wondered how they stay in place. They can be packed on several deep with the bottom layer clinging to hairs on the mother's back. It has been shown that some of the hairs have rows of hooks which catch silk threads that the young spiders lay across the female's back. The spiderlings cling both to the hooks on the hairs and the network of threads, while those on the outer layers cling to the bodies of those below them.

The female wolf spider is also often seen dragging an egg cocoon and it certainly shows more interest in its offspring than most other species. However, maternal care does not extend to those unfortunates that fall off and they are abandoned.

In the afternoon, while I was photographing a family of about ten kangaroos, I heard a great din further up the hill. Turning around I observed a fight between two large male kangaroos. There was much bellowing, fisticuffs and grunting. I had heard that kangaroos stand up like boxers when fighting, and this is certainly true, but they also indulge in kicking. They don't seem to damage each other much even though they have powerful claws. The tussle went on for perhaps five minutes and then the kangaroos slowly hopped down the hill, one following the other — the victor and the vanquished — back towards the family.

I noticed a number of cats today — domestic cats, which have escaped and multiplied. This would not be a cause for concern if they just concentrated on rabbits (and it has been shown that rabbits are the most important item in their diet), but anyone with an interest in wildlife or conservation must be worried about the number of cats in our bush now. They are most effective predators and without doubt are having a detrimental effect on bird life and possibly on some of the smaller marsupials.

Rabbits are also prolific again and this often happens in dry periods. They don't like long grass and droughts provide the right sort of conditions of low grass and bare ground which seem to stimulate their breeding rate.

Wolf spider (Lycosidae).

Late winter afternoon above Towong.

JULY 16

Rain has arrived at last and has that 'set in' quality which you expect in winter. Let's hope it breaks this long dry spell.

We are in a rain shadow here, formed by the Burrowa massif and the Koetong Plateau. Mount Burrowa lies directly to the west, across the valleys of the Tooma and the Murray, and rises to over 1400 metres, its bulk forcing the moisture-laden westerlies upwards, to cool and shed rain. Our rainfall, though good (750 millimetres a year), is thus substantially lower and somewhat less reliable than that of the ranges to the west.

To the east the country beyond the property also rises sharply and the rainfall increases again with elevation. Even within the property we think the average yearly rainfall varies. A rain gauge we kept for several years in one of the higher paddocks tended to confirm this.

The only discordant note from the farm this month was the news that wild dogs are on the rampage again. We recently lost quite a few more sheep and some lambs. So far this year five dogs have been killed, mainly on the boundary of the property. The difficulty at this time of year is that once they start killing live sheep, dogs show little interest in poison baits. I guess the cold weather has brought them out of the park, where game could be fairly hard to find at the moment.

There are a lot of ibis feeding on the Murray and Tooma flats. The Sacred or White Ibis are elegant birds, and with their characteristic shape are easy to identify, even from a distance. They are cosmopolitan, occurring through Africa and the Middle East and into South-East Asia and Papua New Guinea. Today they were feeding with the closely allied Straw-necked Ibis. The latter are peculiarly Australian, rarely straying outside the boundaries of the continent. Like so many of the common birds in this area, they have been encouraged by land development and are favoured by the irrigation schemes and dams. Not only do they nest in swampy country but the insects, crustaceans, worms and reptiles they eat are often associated with partly submerged and submerged areas.

Many magpies are around now. There is a family—about twenty in number—living on a stretch of the Bringenbrong road, several miles from the farm. You can see their nests quite readily in the bare willows along the road. From my boyhood memories I think August is the month magpies become aggressive, swooping on the unwary who get too close to their nests.

Late in the afternoon the sun's rays finally pierced the dark storm clouds, illuminating the dam at the foot of the property. Immediately behind me, to the south, they highlighted the emerald winter grass, at the same time catching the soft blue of the more distant hills. Midway between these two contrasting scenes was a perfectly formed rainbow.

White Ibis near Bringenbrong.

OPPOSITE PAGE ABOVE:
Sun piercing storm clouds above the Tooma River.

OPPOSITE PAGE BELOW:
Wintry scene in Taylors Creek valley.

Magpie and Pied Currawong share a cold perch.

JULY 17

Still more rain overnight. The tracks are now very muddy indeed and it is only just possible to reach the cottage by car.

I left for the snow country at about 9.15 this morning. Light snow began to fall shortly after the turn off to Cabramurra. I stopped further up the road to photograph some scribble gum. The 'scribble' is the remains of the excavation dug by the caterpillar of a small moth which attacks the tree at the junction of new and old wood before the bark is shed. It leaves a dark labyrinthine marking, a curious zigzag pattern almost like some ancient form of script, which stands out on the grey or beige background of the trunk. Scribbles are often associated with White Gum (*E. rossii*) which just reaches as far south as here. However, I collected leaves and fruit of this tree and took them home for identification and discovered that it was snow gum (*E. pauciflora*). Strangely, nearer the snow-line where this snow gum species is dominant, the mining caterpillar does not, in my experience, commonly attack trees.

I spent time today at Cabramurra attempting to take a photograph of an old snow gum in the mist. The sun appeared as a faint white disc, creating an interesting light effect.

Snow gums have been shown to have an influence on precipitation. They don't affect rainfall as such, but in windy conditions the fine droplets in mists are

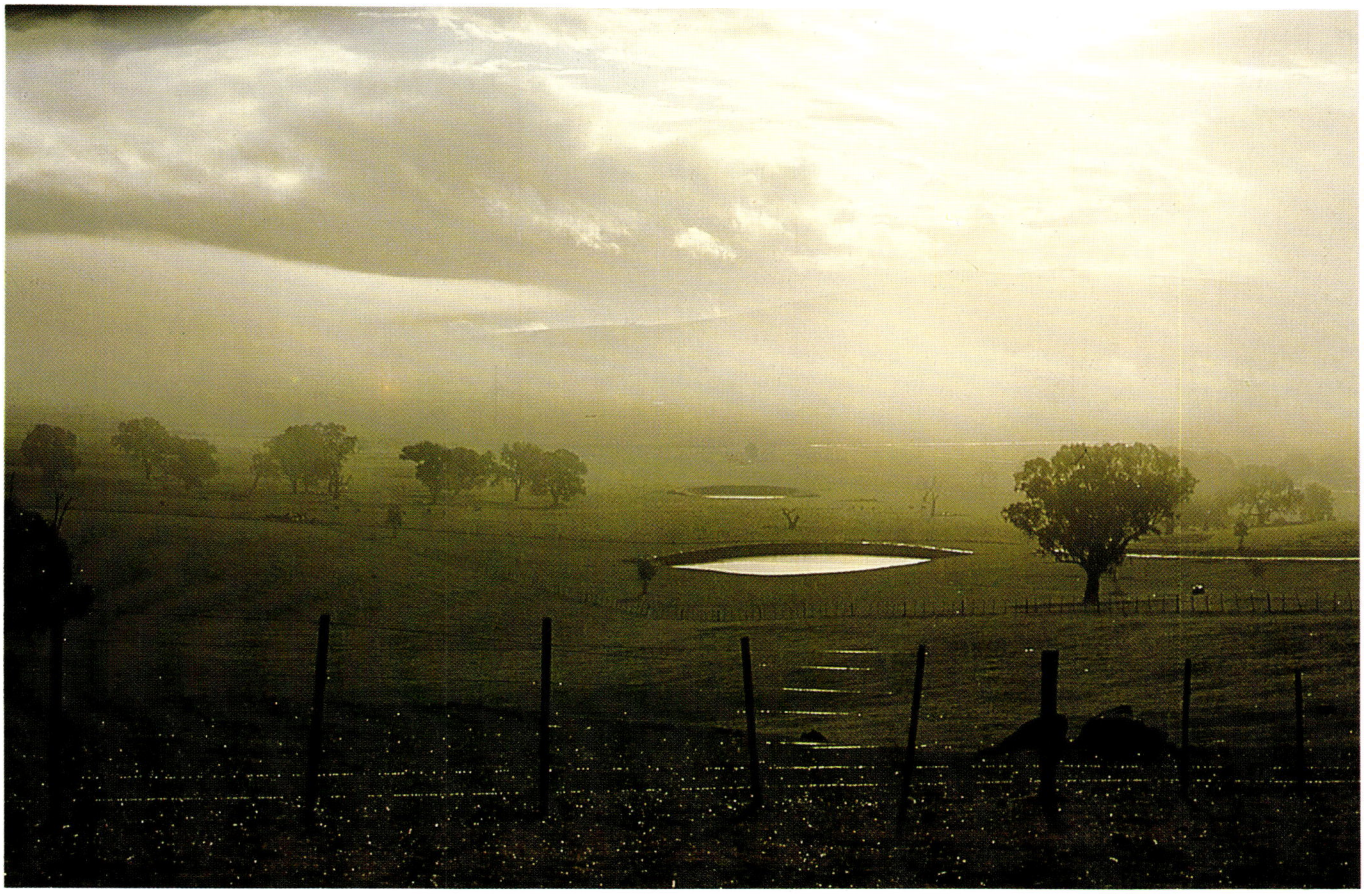

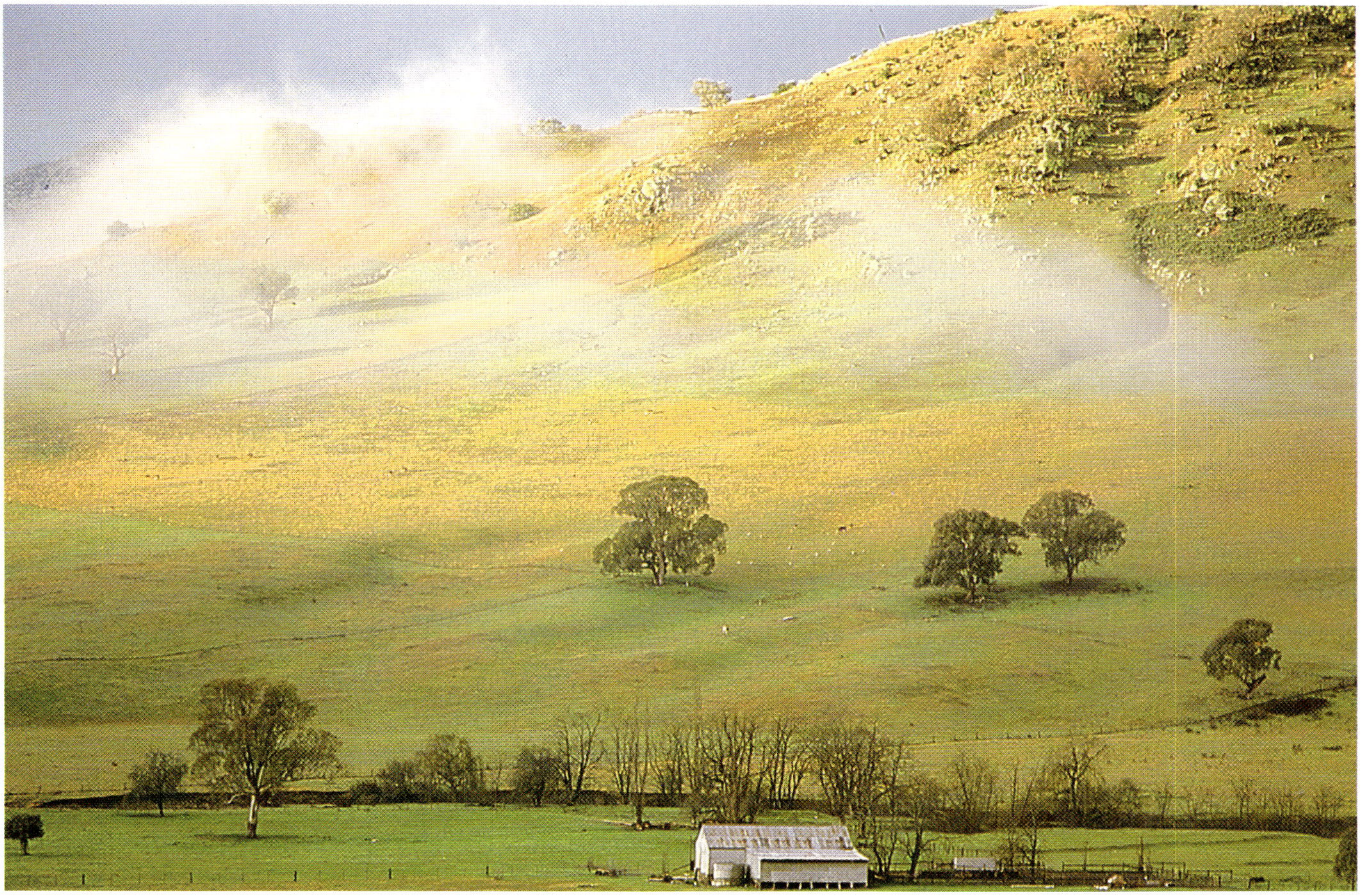

Dead snow gums, Cabramurra.

Caterpillar scribbles on snow gum (*Eucalyptus pauciflora*).

intercepted by the leaves and branches and eventually run off on to the ground. When freezing mists blow in rime collects on the limbs and foliage. In the experimental work conducted in the Guthega catchment in the 1950s it was shown that interception by snow gums added approximately 50 millimetres annual precipitation at 1500 metres and 125 millimetres at 1800 metres.

The snow gum woodland is not continuous, and occurs in a series of small copses. In between these most of the vegetation, mainly small shrubs, is now covered in snow. There was also snow under the trees themselves but only a little. Research has shown that snow gathers most heavily in relatively small clearings among the snow gums. Large open spaces do not accumulate as much snow as without the protection of trees it is easily blown away and melts more quickly.

The accumulation of snow has considerable practical value since it delays the snow thaw in late spring, providing a more even flow of water. On exposed sites it appears that the optimum relationship between the height of the snow gum and the diameter of the open space is about one to four. In protected situations openings between the snow gum copses measuring ten times the height of the surrounding trees proved the most effective.

On the road up to Cabramurra from Tumut 2 Power Station I noted both willows and poplars planted by the Snowy Mountains Authority. They strike an alien note among the native trees and shrubs but serve a definite purpose—the prevention of erosion. Both species appear to have done quite well at this level.

The issue of soil erosion in the Snowy Mountains was a very real one thirty years ago. Although grazing was seen as the biggest problem, the engineering works of the Snowy scheme were more obvious and soil and vegetation disturbance more radical. Not surprisingly the Snowy Mountains Authority was both sensitive and responsive to the issue and much effort was expended on the restoration of roads, dam banks, cuttings and building sites, using whatever species of plants could be established and endure the sub-zero winter temperatures. In most situations only exotics encouraged by fertilisers and artificial mulching were successful. Today many of the turfed road embankments of introduced grasses have been replaced naturally by native species but the poplars and willows remain as monuments to the extensive restoration programme.

I was a little disappointed with my first visit to the snow. The weather was dull and misty and it was impossible to obtain any sweeping view of the mountains. Hopefully I will strike a good day later in the week.

JULY 18

This morning visibility was good at ground level near the cottage, but there was thick cloud at 150 metres. I disregarded the forecast of a cloudy day and decided to take the risk and visit the snow again.

On the way the extensive willow plantings along the Tooma River once more caught my eye. They are very distinctive, particularly during May when the leaves turn gold and define the banks. One of the reasons they were planted was concern about erosion of the stream banks. The willow is an ideal plant to withstand the ravages of a fast running river or stream: the roots are particularly well adapted to wet, poorly aerated soil, and willows can be struck easily from cuttings and proliferate rapidly.

Early morning among the willows, Tooma River.

On the other hand, some willows have grown too large and have tended to block the smaller streams. Even as early as 1902 an official New South Wales government investigation into the deposition of silt over the rich Tooma flats concluded that the problem had more to do with the willows that had developed into a 'perfect jungle' rather than the mining operations at Tumbarumba, though these had also contributed to the level of silt in the stream.

The origin of many willows in the Upper Murray is said to be cuttings from a tree that grew alongside Napoleon's grave on St Helena. William Balcombe, in whose house Napoleon temporarily resided, brought them with him when he came to Australia to settle. He planted some of the cuttings at Kenmore near Goulburn, and further cuttings were taken from these willows to Stuckey's station, at Willie Ploma, near Gundagai. Many of the original Upper Murray settlers passed through Willie Ploma and took willow cuttings to plant around their new homes to provide softer and greener foliage than that offered by the harsh new landscape.

Just before the turn off to Kiandra and the snow-fields I saw three Emus feeding among grass tussocks. They are essentially vegetarian and eat a lot of young shoots and the seeds of various fruits. You find them throughout this district and it is obvious that they move widely in search of food. During the breeding period in winter, when the male is engaged for about eight weeks in incubating the eggs, they tend to remain in the one location.

On the way to Cabramurra the gorge of the Tumut River unfolds dramatically, dropping 700 metres as you leave the gentle undulating forest country on the Bago Plateau. Snaking down the narrow dirt road to the impounded waters of the Talbingo Dam, it reminded me of an account I'd read by Lieutenant-Colonel Freeling, who was sent by the South Australian government in 1860 to report on the best routes available to the gold-fields at Lobs Hole and Kiandra. Lobs Hole is at the junction of the Tumut and Yarrangobilly rivers and is now covered by the waters of the Talbingo Dam. Freeling wrote that

> The descent into Lobbs Hole, where there is rough accommodation in two shanties, is for a distance of about two miles so steep that horses have to slide down nearly on their haunches, and persons on foot must observe the utmost caution in descending.

I returned to the same spot I visited yesterday, Kings Cross, east of Cabramurra, a small rise which provides 360° views. The area overlooks the Dargals and the prominent peak Jagungal, which rivals Kosciusko in this landscape. It is 167 metres lower but its isolation makes it stand out as no other mountain does.

A few kilometres to the east I came to the broad, open, treeless valleys which characterise the Kiandra district. They are cold air drainage basins, sometimes called frost hollows. During the evening cooler and heavier air flows down into the basins, rising as the morning sun warms the atmosphere. Ground conditions in the hollows are too cold to allow germination and development of snow gums.

You can still clearly see the sluice race lines cut neatly along the contours of the valleys, reminders of one of the most short-lived and spectacular gold-rushes in Australian history. Gold was discovered at Kiandra in November 1859, the rush began in January 1860 and by April of that year there were 10 000 miners on the Kiandra field. One report claimed that at its peak 500 ounces of gold were won daily from an area of about a square kilometre. The more easily found gold was soon exhausted, however, and by March 1861 only 380 miners remained. Mining continued until 1905 with up to 200 men employed on larger operations which involved hydraulic sluicing and dredging. The Kiandra field yielded 172 000 ounces of gold and almost half of this was found in the fifteen months that the rush lasted.

Some reef gold was found at Kiandra but the major source was alluvial. At first this was mainly from the sand and gravel at the base of the existing streams but later from the beds of ancient streams overlain with basalt. These were the so called 'deep leads' and with larger amounts of overburden surface sluicing proved inadequate and was replaced by hydraulic sluicing involving the construction of storage dams.

Pattern of ice in pond near Kiandra.

The solitary peak of Mount Jagungal, viewed from near Cabramurra.

Access to the gold-field was very difficult because of the rugged terrain, with the huge gorge of the Tumut River to the west and the formidable 'Talbingo Hill' to negotiate on the main route in from Tumut to the north. A way in to the field from Victoria, avoiding the precipitous and dissected country where the Tumut leaves the Murray Plateau, was established by a surveyor, Ligar, and was called Ligars Route. A track was also established from Eden in the east. Despite the great difficulties of terrain and climate, many miners braved the winter snow, mostly in tents, and there were still 4000 on the field in August 1860.

Coloured lichens and alga on fence post. The red-brown alga near the base favours the southern side, whereas the yellow lichen grows most prolifically on the eastern side.

JULY 19

Dull and wet this morning with rain forecast for today and tomorrow. There were forty-four fogs recorded in Corryong for the two months May–June. This seems to be something of a record, even for this district.

I noted colourful lichens and an alga growing on the fence posts along the side of the road: vivid ochre-red alga (*Treatepohlia aurantiaca*) and yellow and green species of lichens. Some of them are sensitive to temperature or light, perhaps both, because the cool southern faces of the posts were invariably red, almost as though painted, whereas the yellow lichen (*Chrysothrix candelaris*) dominated the eastern side, which catches the morning sun.

JULY 20

This morning while having breakfast I looked out the window to see a pair of Double-banded Finches feeding on the bird tray I've erected. According to one reference book I have they don't occur south of mid-New South Wales. However, on checking this further I have discovered that more recent observations have extended their known range and there have been sightings of them at Albury, but to date not on the south side of the Murray, so far as I know.

Despite the forecast, which suggested rain and heavy snow, I decided to go again to the high country. I broke out of the mist at about 600 metres, near the turn off to Kiandra. In fact it turned out to be an excellent day, particularly for photography.

Snow gum (*Eucalyptus pauciflora*) near Cabramurra.

Late afternoon, Kings Cross, near Cabramurra.

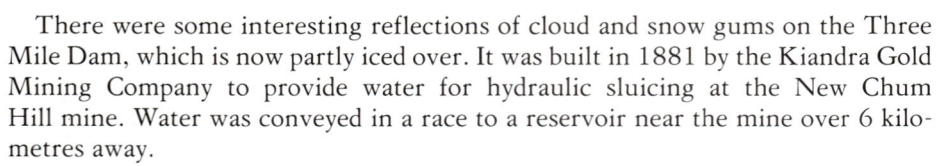

There were some interesting reflections of cloud and snow gums on the Three Mile Dam, which is now partly iced over. It was built in 1881 by the Kiandra Gold Mining Company to provide water for hydraulic sluicing at the New Chum Hill mine. Water was conveyed in a race to a reservoir near the mine over 6 kilometres away.

Three Mile Dam, between Cabramurra and Kiandra, built in 1881 to provide water for hydraulic sluicing.

JULY 21

In thick fog I set off early this morning to Mount Elliott. I stopped briefly to photograph the sun streaming through the mist and forming a silhouette of an old pine tree. Just near by was an abandoned water-filled shaft driven horizontally into the hillside. This is near the site of the 'New Chum' reef, the 'big' find of the Mount Elliott field which was made by the McInnes family in 1894. The mine yielded £100 000 of gold in those days and its discovery initiated the Mount Elliott rush. Eventually water prevented further operation of the mine.

Old pine tree near the site of 'New Chum' reef, Mount Elliott.

The summit of Mount Elliott is one of the major vantage points in this district. Looking out over the cloud to the east I could just see the peak of Mount Mitta Mitta, the point over which aircraft change direction between Melbourne and Sydney. In the opposite direction the deep blue, white-capped Kosciusko massif dominated the scene; I would like to be here in a week or two to photograph the contrast between these colours and the Silver Wattles which are just about to burst into flower.

There is a thicket of Silver Wattles on the eastern side of the summit and they are spreading out across the clearing. I wondered whether the seed could have been dispersed by ants. It is known that ants play an important part in the life cycle of many plants, including Silver Wattles. The seeds have specialised structures (elaisomes) which attract ants. This relationship is highly developed in Australia and 7 per cent of the Australian flora is thought to rely on ants for seed dispersal.

Flock of Yellow-tailed Black Cockatoos.

In the afternoon as I went looking for kangaroos in one of the high paddocks next to the National Park boundary I heard a flock of black cockatoos approaching. They flew directly overhead in neat formation, the creamy-yellow markings on their tail feathers clearly visible.

I noticed a rabbit with myxomatosis. There is not much evidence of the disease this early as it is mostly associated with the warmer months. The mosquito and the European Rabbit Flea are vectors of myxomatosis, carrying the virus on their mouth parts from rabbit to rabbit. However, whereas the mosquito is dependent on the season and appears in the warmer months, particularly in humid weather, the rabbit flea is active over a much longer period. Not only does this extend the infective period of myxomatosis, but it highlights the advantage of a winter outbreak of the disease when the rabbit is more susceptible. Even older field strains which are only partially effective in the warmer months can kill a rabbit during winter. Moreover, a much bigger dent is made in the population if there is a winter outbreak, before the numbers begin to build up in the spring.

The rabbit flea was introduced from Europe during the 1970s. It was first carefully tested and shown not to infect native animals. It is completely dependent on the life cycle of the rabbit. The reproductive cycle in the flea can only be triggered after a meal on the blood of a pregnant rabbit or doe. In fact the whole life cycle of the rabbit flea is regulated by the hormones secreted by the rabbit during pregnancy. The fleas sense the end of the pregnancy, migrate to the ears and cling strongly to the edges. Shortly after the young rabbits are born the fleas leave the female and infest the progeny, extracting a substance required for their successful mating. Eggs are laid in the rabbit's nest and when the larvae emerge they feed on blood excreted by adult fleas. Eventually the larvae pupate, the adults appearing in fifteen to a hundred days, depending on conditions. But the flea adjusts to the number of generations of rabbits. There can be up to five or six generations in one year and the fleas modify their life cycle accordingly.

JULY 22

Today I noted large growths on a number of Narrow-leaved Peppermints and I've seen them often enough on red gums. They are grotesque and may be a metre or more in circumference. The full story behind these curious growths is still to be unravelled. It is believed they are often initiated by fire. When a tree is defoliated by fire, or the limbs removed by wind storms, many vigorous shoots may develop from a single bud strand in the live bark. In some instances the lower stems of the shoots join up to form an epicormic knob, and sometimes a group of these develop close together. When the tree recovers the knobs are suppressed and the trunk grows around them, creating the outgrowths.

Recent evidence suggests that in some instances insects may be involved, infecting the epicormic shoots and stimulating uncontrolled proliferation of plant cells in a fashion similar to cancer in animals.

I was concerned to find that one of the red gums we planted at the farm several years ago looked dead. I went over to examine it and found some sawfly larvae on one leaf, but the leaves were discoloured and mostly dead and I think the cause may be a sap-sucking lerp insect. The immature stage of this strangely named insect forms a scale, or lerp. Many of these scales, especially when viewed under a magnifying glass, form an unusual variety of shapes, some like sea shells, often intricately decorated. The scale protects the soft-bodied immature stages of the lerp against desiccation. The infestation itself does not usually kill the tree but may be accompanied by attacks from other types of insects such as Christmas beetles, sawflies and various moth caterpillars. Successive outbreaks can result in the death of farm trees and are now recognised as a major problem in south-eastern Australia. Many eucalypt species recover rapidly from insect defoliation by the production of epicormic shoots, but their capacity to do this is limited and if the new shoots keep being eaten the tree will die.

Lerp insect (Psyllidae) and scales on eucalypt leaf.

Old red gum (*Eucalyptus blakelyi*); note huge growth on the trunk.

Eucalypts have probably always been subjected to relatively high levels of insect infestation. Near Canberra it has been shown that insect damage results in the loss of 70 per cent of the herbage from red gums (*E. blakelyi*), including those growing in natural uncleared woodland, contrasting with the much lesser damage recorded for trees found in temperate forests in the northern hemisphere.

August 11

There have been several days of heavy rain this month, following the 100 millimetres already received in July. What a change after such a dry autumn and early winter. Most of the tracks around the property are still very muddy and it is easy to get bogged, and the waterfalls are running strongly.

I saw a Magpie-lark attacking a Brown Hawk, a seemingly unequal contest, but the Magpie-lark was very aggressive and soon drove the hawk away. Presumably it is concerned that the Brown Hawk may locate its nest and steal its young.

Looking around the pastures I was struck by the fact that there wasn't much clover. The most likely reason is that we didn't have late autumn rains. The sub clover germinated because of the good rain in late March, but without follow-up rain much of it seems to have died. This would not be due to a phosphate deficiency because superphosphate has been spread consistently on the paddocks over the years and most of the property has had at least 2 tonnes per hectare.

Magpie-lark attacking Brown Hawk.

It is possible, however, that the lack of clover could be due to a trace element deficiency. Molybdenum is required on some soils, particularly on hill country, for the effective functioning of bacteria, or rhizobia, which form nodules on the roots of clover and fix atmospheric nitrogen, making it available to the clover and ultimately to the soil and associated grasses. Without molybdenum rhizobia are not effective. Perhaps the most remarkable fact about molybdenum is the tiny amount needed to produce a result, as little as 50 grams per hectare being required to maintain an adequate supply in the soil for several years.

We did pasture trials with trace elements in one hill paddock many years ago and could obtain no indication of a lack of molybdenum, and clover seems to do well here. However, since molybdenum deficiency has been noted in some parts of the district it is possible that it could occur in other paddocks on this property.

AUGUST 12

Another very wet day. We stayed indoors most of the morning and went eventually to Welumba Creek for a picnic. We finished up having our lunch under a tree and all getting drenched.

We were interrupted by a small but noisy flock of black cockatoos. You can readily identify them because of their lumbering flight and their size, larger than any other cockatoo in south-east Australia. They have a wing span of about 60 centimetres and a long tail which gives them a characteristic outline when they are in the air. I guess they've come down from the higher country looking for seeds, and there are lots in the Welumba Creek area that would suit them. They eat wattle pods, and seeds of banksias and Black Cypress-pines, which occur on the hills on the north-eastern side of the creek.

I also saw a case moth. The caterpillar had cut off even lengths of small twigs to form a stout protective case. In many Australian species of case moths the female is wingless. It mates while still within the case and in some species the young caterpillars find their way out on silken threads, a means by which they can be transported by wind.

Next to where we picnicked was a large Blackwood tree, a member of the wattle family. There are over 650 species of wattles in Australia and this represents around two-thirds of the known species in the world.

Most wattles, including the Blackwood, have phyllodes, or modified leaves, formed from the flattened leaf stalk, or petiole. Initially all the wattles with phyllodes start out with a true leaf, which has a feathery pinnate or bipinnate shape, but in species like Blackwood, the original bipinnate leaves are replaced by a phyllode. What interested me was that you could see on the end of some of the phyllodes the feathery bipinnate immature leaves, almost as though the plant had not yet made up its mind.

The bipinnate leaves are thought to be more effective in 'catching' sunlight. Growing as it often does in the shade of larger trees, the young Blackwood presumably makes use of its juvenile foliage to grow more rapidly until it reaches a better lit position.

Some of the phyllodes in Australian wattles take on quite remarkable shapes—particularly in the arid regions—and may resemble thorns, needles or daggers. The number of veins on the phyllode is an important diagnostic characteristic used in the classification of the group.

Several species of wattles are now coming into flower. There are six common species in the area: Blackwood, Silver, Dagger, Box-leaf, Varnish and Juniper. In one place the bright yellow of the wattles was offset by the most attractive purple of Coral-pea and the lighter lilac of another pea flower, Lance-leaf Hovea.

Blackwood (*Acacia melanoxylon*) phyllode with immature bipinnate leaf.

Juniper Wattle (*Acacia ulicifolia*).

AUGUST 13

Again a very dull day and not at all favourable for picnicking or sightseeing.

I inspected a paddock containing perhaps a hundred fine specimens of Red Stringybark. Although there are vast numbers of these trees in the dry foothill forests, they are becoming increasingly rare as paddock trees. The problem is that in winter cattle develop a craving for the bark; it comes away easily in large sheets from the trunk and the trees are sensitive to its removal. The result is that huge numbers of these trees and other species of stringybark have been killed in recent years in Victoria and New South Wales. The problem is encouraged by pasture improvement with the consequent high proportion of nutritious clovers and grasses and a relatively low fibre content. In our experience cattle do not tackle Red Stringybarks growing on unimproved pasture land. Cattle cause most damage in July and August when paddock feed is at its lowest ebb, and what is available has a high moisture content and little fibre. The simplest explanation of the craving is that as ruminant animals, the cattle need roughage and hence go for the bark. Another explanation is that salt deficiency is the cause and there is evidence that a craving for magnesium could also be involved.

The only answer to this problem, short of feeding out extra quantities of hay (and possibly salt and magnesium), is to protect trees either individually with wire guards, which we have done to some extent, or to exclude whole sections of trees from a paddock. I was pleased to see that the wire guards we strung on many trees ten years ago are still providing effective protection, and that even unprotected trees, although damaged, are still in good health because of our relatively low stocking rate in this paddock.

This season's lack of clover does not bode well for the spring; even if we have good rain the pastures and hay are likely to be predominantly grassy and of low quality. On the other hand, we are unlikely to face the menace of bloat. In some years bloat can be very worrying, striking suddenly and then disappearing almost as quickly. It is always associated with pastures rich in clover and is usually worst in late September and early October. Caused by a foam which blocks the rumen it can only effectively be controlled on an extensive grazing property by avoiding clover-dominant paddocks or by feeding cattle blocks which contain some detergent material to break down the foam. But to do this successfully requires anticipation of the problem, because it occurs so suddenly.

AUGUST 14

We drove up to Cabramurra today to take the children to the snow. Several small Pied Cormorants were diving in the reservoir near Sue City and I was intrigued to see that one came up about 50 metres from the place it dived. They obviously swim at great speed under water.

At Cabramurra there was a much heavier covering of snow than I'd experienced last month. A small flock of brilliantly coloured Crimson Rosellas looked spectacular against the white snow. It made me wonder what they'd find to feed on in this barren environment. There doesn't seem to be much in the way of fruit or berries.

Pied Currawong in the snow.

SEPTEMBER 9

There has been a great change since I was here last. It was a miserable snow season to begin with—poor falls and disappointing skiing conditions—but it has finished with a 'rush'. The whole top of the main range is covered like a giant cream cake.

What a picture the Upper Murray is at this time of year. Looking south-east from above the Towong Bridge there is a great expanse of green spring pasture, the early morning blue of the meandering river contrasting with patches of yellow wattle along the banks and a backdrop of snow-covered ranges.

I paid a visit to Welumba Creek. Many wild flowers are out, and I found a few of the small spider orchids known as Pink Fingers. They are not always pink, sometimes white or light red, but you always find them early in spring and they are one of the commonest of the terrestrial orchids in southern Australia.

Early spring: the Murray River above Towong Bridge.

I also found some Pale Sundews. They were not yet in flower, but the curious leaves with their bright red tentacles made them obvious. Australia is particularly well endowed with sundews and all are insectiverous. This group stood about 30 centimetres high, with the sun behind them, shining through the minute red glands on the leaves. The flowers are most often white, and some species are prostrate, barely 2 centimetres high. Sundews frequent badly drained, infertile soil, and are able to utilise these poor habitats because they can capture insects and so obtain nutrients to augment the low supplies in the soil.

Looking closely at the leaves I could see that the large tentacles are on the outside of the leaf while those in the centre are considerably shorter. The various remains of insects were nearly always in the centre and it is the glands on the shorter hairs which are useful in digestion. When an insect alights on the centre of a leaf the mucilage in the glands is usually strong enough to hold it. If the captive struggles violently it touches tentacles immediately around it which move inwards to enfold it. By some process not fully understood the struggling movements set off a reaction through the leaf blade and the outside longer tentacles also curl inwards and assist in securing the prey.

The leaves of the Pale Sundew (*Drosera peltata*).

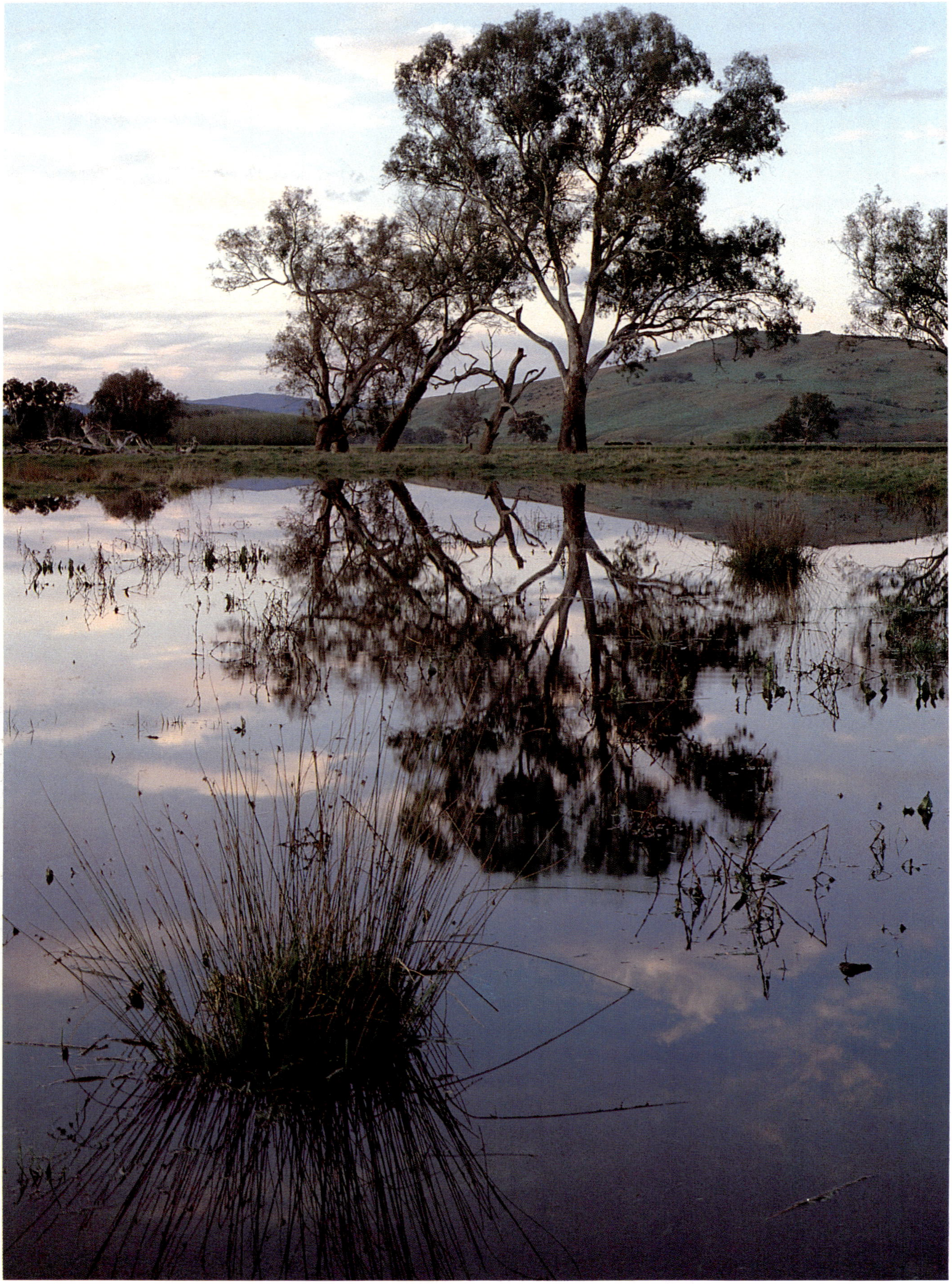

The tentacles with their glandular tips have a three-fold function. They secrete mucilage to catch the insect, enzymes to digest it, possibly with the help of other small hairs on the leaf, and they absorb the liquid remains of the insect.

It has been shown that the sundew can be 'tricked' in so far as it will envelop a breadcrumb placed on the leaf. However, eventually it 'realises' that this is not suitable food and the tentacles release their hold within twenty-four hours.

It is not known what attracts insects to the sundew leaf. The glands do not secrete nectar or its equivalent, but it is possible that the insects are lured by the glands' glistening appearance which may give the impression of nectar, or it may be that some subtle undetected scent is emitted by the plant. Whatever the means of attraction it is effective because every plant examined had insects on the leaves.

The Coral-pea, which is very plentiful here, is always one of the harbingers of spring. It is a very vivid colour, usually a deep mauve with a background of coarse dark green leaves. Frequently you will see it growing on rocks or dead tree trunks, and often it cascades over the bank where a road has been cut through the forest.

OPPOSITE PAGE:
Flooded pastureland near Towong.

SEPTEMBER 12

It rained heavily during the night. We have had over 125 millimetres in August and 50 millimetres to date in September, and the season has turned right around. Following on from the long dry spell we had during winter, the heavy rain emphasises the fickleness of the climate in this part of the world.

The river flats of the Murray and the Tooma this morning were flooded. Groves of red gums, which for months have been 'high and dry', are now surrounded by water. I saw two White Egrets among one group of gums, birds and trees casting sharply defined reflections in the perfectly still water.

White Egrets.

What primarily determines stream flow and flooding is, of course, the rate of precipitation in the catchment. You need a situation where the soils are well saturated and the recent rains have assured that. Last night's sharp hard falls were sufficient to allow much of it to run down into the streams. The other factor determining the rate at which a river rises is the amount of cleared country in the catchment. Where the hill country has been denuded the rate of run-off is much higher than from forested country.

I checked the stream flows for the Upper Murray. Records have been kept since 1890 at Jingellic downstream from here. The maximum annual discharge was just under 6·5 million megalitres in 1956, compared with the average of 2·5 million and the mere 0·75 million megalitres recorded in the drought of 1902, clearly demonstrating the huge seasonal variations that can occur in the head waters of Australia's principal river.

As the flood waters rose around me and began to cross the road to Tooma, I was interested to observe a group of rove beetles (Staphylinidae) comfortably swimming on the surface. It almost seems as if this species has been provided with the bouyancy and water-repellant cuticle to cope with floods. These were dark beetles, about 2·5 centimetres long with short wing cases which left the abdominal segments quite visible. They didn't really look like beetles at all but were more like earwigs, without the rear pincers of those creatures. The abundance of insects flushed out of the grass by the rising water would account for much of the interest of the great number of water birds present.

Early evening along the flood waters near Towong.

SEPTEMBER 13

I went up into the National Park and had a look along the transmission line. There are many flowers, including the Red-stem Wattle. Its individual flower balls are almost as large as those of Golden Wattle.

I also found the caterpillars of a skipper butterfly (*Hesperilla donnysa*), identified at this stage primarily by distinctive markings on the head. The caterpillar of this particular group of butterflies feeds on sword grass and seeks protection by joining a number of leaves together in the form of a cylinder. This provides an excellent shelter and when the caterpillar is inside it seems impregnable from enemies. Nevertheless, a surprising number are parasitised by wasps and flies and one can only assume that they are discovered when they emerge from the shelter during the evening.

Red-stem Wattle (*Acacia rubida*).

The coral heath was in full flower and although it was a little early in the season, at this elevation of 700 metres, for the heath to attract many insects, there were a few hover-flies. They are an intriguing insect. So quickly do their wings move that they give the appearance of being suspended in mid-air without support. Hover-flies can remain completely stationary above flowers and yet can move with great speed to one side, or drop or rise without warning and then resume this perfectly 'motionless' poise. They are among the fastest of all insects.

Hover-flies are also splendid 'mimics'. Their colour patterns, with yellow or orange bands and the constricted waist, give the appearance of some of the stinging wasps, presumably giving them extra protection.

Today I stayed much later than usual on the transmission line and I'm glad I did. Just before the sun set the wallabies came out from the forest on either side. I guess they've seen cars before because they were fairly tame, as long as I didn't get out. These were Red-necked Wallabies, variously and subtly coloured but usually red-brown on the neck and face, merging into shades of grey on the body.

Female Eastern Grey Kangaroo, joey in pouch, with the larger brown coloured male.

Lower down there were a few Eastern Grey Kangaroos, more wary than the wallabies, but again much tamer than those around the farm. A male and a female looked at the car curiously without showing any signs of moving. I rolled down the passenger's window, crawled across the front seat and began to photograph them. During this process a joey popped out from the mother's pouch to have a look, too.

The young kangaroo begins to leave the pouch at about nine months for short intervals and permanently at eleven months. It is immediately replaced by a newborn kangaroo developed from a blastocyte whose development has been held over while the joey is being suckled. The joey returns to the pouch at intervals and attaches to the same teat it has used during its period in the pouch. Most remarkable of all, the female produces different quality milk for her two offspring.

What struck me as being unusual was the quite reddish colour of the male. I've noticed that some of the big male Eastern Grey Kangaroos are brown rather than grey, but this was almost the colour of a Red Kangaroo.

Tonight I found a hawk moth on the verandah. It is a quite common species (*H. scrofa*), with brown leaf-like forewings and bright orange hindwings. The forewings have a vivid white line on them as do the antennae. The whole appearance of the insect is one of great streamlining and speed. Hawk moths are very fast and records suggest that they can fly at least 15 metres per second (over 50 kilometres per hour).

Hawk moth (*Hippotion scrofa*).

September 14

Much to my disappointment the Yellow-tailed Thornbills have abandoned their nest in the jasmine on the front verandah. Perhaps one of the wild cats discovered it and disturbed them. It is a surprisingly large nest for so small a bird, located deep within a thicket of jasmine, and the birds have made liberal use of the wool lying about in the paddocks in its construction.

Later on this morning I took photographs of the Welumba Creek waterfall, which is much more turbulent than it was during the dry winter months. However, the flood waters on the Murray and Tooma have gone down today. Flooding is transient in this part of the world. The rivers come up one day, flood streamside paddocks and even cut bridges, then subside again.

I was struck by the appearance of the Box Wattle, its glowing orange-yellow flowers contrasting with the jet black trunks of the Red Stringybarks, burnt a few years ago. The understorey is dominated by Violet Kunzea, a small shrub about a metre high, with minute grey-green leaves and buds soon to burst into mauve flowers. In this situation the kunzea is almost acting like a weed, filling a niche formed by recurrent firing and past grazing.

Near by were some Slender Rice-flowers. They belong to the Thymelaeaceae family, which includes the daphne. Most of them contain neuro-muscular poisons, one of which is daphnin. Poisoning from rice-flowers is not often reported in

Box Wattle (*Acacia buxifolia*) and Red Stringybark (*Eucalyptus macrorhyncha*).

Slender Rice-flower (*Pimelea linifolia*).

Welumba Creek waterfall.

Australia, although several species are suspected of being poisonous to stock. One of the more interesting things about rice-flowers is that the fibre of some species is extraordinarily strong and was used by Aborigines to make nets. It must have rated highly among the useful plants of the Australian bushland.

I also noted three species of beard-heath. At first glance it is a little like common heath, except for the beard or hairy nature of the flower. It is an intricate flower and you need a magnifying glass to show up the delicate structure of the throat.

As I was taking the photographs a bull ant passed me carrying the remains of an insect back to its nest. (I saw a black one earlier feeding on honeydew from scale insects on a leaf.) When I'd finished my photography I attempted to track it down and, after some time, located the nest. Frankly it didn't take much locating once I was in the right area because several savage red ants came out to see what was going on. A large nest may contain a thousand bull ants and is fairly deep, going down as far as 2 metres, with galleries and tunnels arranged in a more or less vertical configuration.

The bull ant is a member of a group of ants which is almost exclusively indigenous to Australia. Only one species occurs outside Australia, in New Caledonia. Bull ants are extraordinarily fierce and they kept attacking as I tried to photograph them. They have a potent sting but fortunately I didn't experience one today—although I've been stung many times in the past. It seems that the ferocity of these ants varies from species to species. This particular red variety is one of the most aggressive that I have encountered. The sting of the bull ant is not barbed and does not remain in the victim, unlike that of the bee, and the ant is capable of repeating the dose. Strangely, the grasp of the huge mandibles or jaws of the bull ant workers is quite gentle and they do not bite, unlike many other ants.

One of the more unusual aspects of bull ants, which are very primitive ants, is that the queen leaves the nest at night to forage and supplement the food supply of her progeny. In more highly developed families of ants the young are nurtured by the more or less immobile queen from metabolised fat and other material within her body.

Coming home tonight there was a great sunset. The road leading out of Welumba Creek heads due west and is perfect for viewing sunsets, which are often spectacular here either before or after a rainstorm. There was a lot of rain at Albury today—145 kilometres down river—but it didn't get up this far. Nevertheless the clouds to the west, above the purple outline of Mount Burrowa and Pine Mountain, caught the dying rays of the sun, becoming brilliant streaks of red in an otherwise dark and threatening sky.

Beard-heath (*Leucopogon biflorus*).

Bull ant (*Myrmecia sp.*).

Stormy sunset, Welumba Creek.

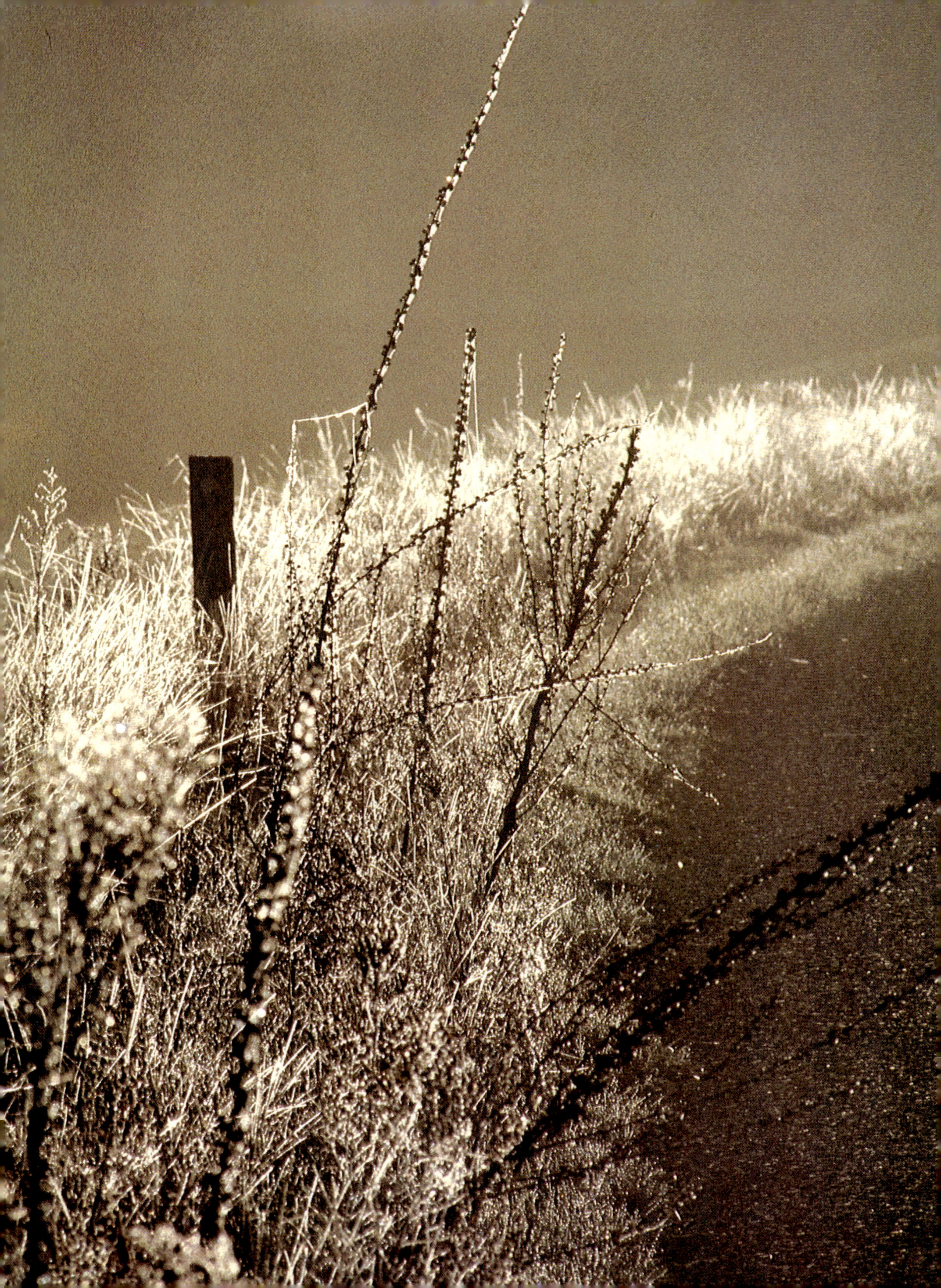

Sun breaking through.

SEPTEMBER 15

Probably one of our last frosts for the season this morning. According to the records there aren't many frosts after mid-September and in fact, it should stay frost free now until the end of April.

Patches of mist clung to the water in the billabongs in the soft light. Even as late as 8.30 a.m. the sun had not really broken through the mist and there were some extraordinary reflections in the still water. Every now and then an insect striking the mirror-like surface would send tiny concentric waves out towards the banks.

On the way back to the farm I stopped on the crest of the road and looked out across the Tooma flats and down the Murray towards Tintaldra. The flood waters have receded though the billabongs are still full, and it is now easy to see how far away from the existing river bed some of the ancient courses must have been.

Several Brown Hawks were circling, looking for rabbits and there will be plenty when myxomatosis starts to take its toll.

OPPOSITE PAGE:
Hopefully the last frost of the season.

Caterpillars of the Imperial White butterfly (*Delias harpalyce*).

Pupae (*Delias harpalyce*).

Imperial White butterfly.

In the afternoon I had a look at the pasture we sowed earlier in the year. The result is very poor and unless we have a prolonged spring I doubt that it will succeed. Two years ago on this same paddock we had a good strike, and until October of that year the pasture looked excellent, but virtually nothing survived the drought. This year the sown grasses and clover are small and sparse, partly because it is dry and partly because we lightly cultivated the area twice before sowing, to level out the very rough surface. It is hill country of naturally low fertility, and the cultivation appears to have turned in the top few more fertile centimetres. In a year of normal autumn and winter rainfall the plants at this stage would be much further advanced and able to cope with the long dry summer.

Usually we sod-seed such areas—and in fact most of our pastures—which provides individual furrows and minimises soil disturbance. The existing pastures are sprayed with the herbicide paraquat, and the seed and fertiliser are sod-seeded directly into the furrows in the desiccated sward. With the right combination of herbicide, seed, fertiliser and equipment, and if your timing is right, this technique is very effective and economical. It takes far less time than conventional cultivation and also greatly reduces the risk of soil erosion on steep land.

In one unsown section, still predominantly native grasses, there were a few plants of Early Nancy. This quaint small lily has a white flower, with a transverse purple band about one third from the base. The Early Nancy is an example of a plant with flowers of different sexes—female, male and bisexual flowers. In my experience it is one of the native plants that has adapted to the clearing of the original forest, where pastures have not been developed. As the name suggests, it is one of the first native plants to flower at the start of spring.

I was pleased to find a group of caterpillars of the Imperial White butterfly on a clump of Drooping Mistletoe. I've been keeping an eye out for this butterfly and its caterpillars for some time; it is one of the most colourful we have in south-eastern Australia. The Imperial White belongs to the genus *Delias*, with white, or black and white upper wing surfaces and a brilliant underside—usually a mixture of red, yellow, black and white.

Like most members of this group, the caterpillars feed on mistletoe. Imperial White larvae are gregarious and rest together on leaves and twigs, spinning a silken web on which they pupate. The pupae can be quite mobile and are able to make sharp twisting movements. If disturbed, the movement of a single pupa can set off a chain reaction among the others on the web. Doubtless this helps protect them against birds. For some reason spring pupae seem mostly to be black, while those of summer and autumn are often bright yellow. Overseas research on related species of butterflies suggests that it is the colour and the direction of light to which the caterpillars are exposed immediately prior to pupation which controls the colour of the pupae.

This paddock we are currently sowing down abuts a forested area we hope to fence off to conserve. It is a southern-facing piece of woodland, dominated largely by Broad-leaved Peppermint but with some Red Stringybark. It has been little affected by grazing over the years and is a good place for wild flowers.

I found three species of greenhood orchids in the forest. One of these, which is a hybrid (*Pterostylis x ingens*) of the Nodding (*P. nutans*) and the Sharp Greenhood (*P. acuminata*), was very common, with groups standing 15 to 25 centimetres high in the grass. I also found the Maroonhood and the Blunt Greenhood. These predominantly green orchids are nearly always found in shaded places.

The pollination mechanisms of greenhoods demonstrate the intricate and specialised relationships that can exist between insects and plants. Greenhoods have a labellum, or tongue, with an appendage at the base which is thought to be an attractant. When the appropriate insect alights on the delicately poised and sensitive labellum in the vicinity of the appendage, the labellum moves upwards and the insect is trapped between it and the column bearing the reproductive parts. The hood or galea is made up of the dorsal sepal and two lateral sepals with translucent areas that act as a 'sky light' to attract the insect as it seeks a way to escape. It moves towards the light and is finally 'directed' through a tunnel formed by the wings of the column, and to escape must push up past the anther bearing the pollen masses, or pollinia, which become attached to it. When the insect visits another greenhood of the same species the process is repeated. The pollinia come in contact with the stigma on the column and cross-pollination is achieved.

As I was photographing the hybrid greenhood a mosquito landed on it, examined it but made no effort to enter. Perhaps the orchid had not reached the stage where it was receptive to the mosquito, or maybe that particular mosquito was not the species responsible for pollination.

Dissected Nodding Greenhood orchid (*Pterostylis nutans*). Hood removed to reveal the sensitive labellum with its appendage and the column bearing the stigma (midway) and the anther at its apex, with yellow pollinia just visible.

A colony of the hybrid greenhood orchid (*Pterostylis x ingens*). Such colonies arise by vegetative reproduction.

OCTOBER 21

The pastures have not improved. Although they are green and appear lush, the lack of clover means they have, as the farmers say, no 'body', and any hay is unlikely to be of good quality, nor will there be a great number of bales.

It is likely that pastures, particularly on the hills, will cut out fairly quickly in the new year. Farmers look to the spring growth to provide the dry standing feed to last through to the next autumn, but it now looks as though this could run out early in February rather than in late March or April next year.

Dawn on the Murray River between Towong Bridge and Farrans Lookout.

On the way to Welumba Creek I flushed out a couple of Red-rumped Parrots. They are always exciting to see, beautifully camouflaged when they are resting in the grass but very colourful in flight. The Red-rumped Parrots seem to disappear early in winter and this is the first time I've seen them again. They would be looking for nesting hollows now, usually in eucalypts and often near water.

I also saw some Red-browed Finches, which I am sure are nesting in the briars along the fence lines, although I've not yet found a nest.

Another group of birds I observed were magpies. These have taken up residence in our front paddock. I've always seen a few here, but there are many more than usual. There was a mixture of adults and young birds, the adults intensely black and white and the immatures slightly smaller and grey rather than black. The adults seemed to be training the younger birds, accompanying them for short distances on to the pasture as they made tentative efforts to fly.

At Welumba Creek I heard the characteristic screeching of Gang-gang Cockatoos, but it took me a while to track them down. I eventually found a pair sitting on the

Sheep grazing spring pastures above the Tooma River.

Gang-gang cockatoos, Welumba Creek.

branch of a dead tree near the track. Of all the cockatoos in the wild the Gang-gang is easily the most approachable. I had no difficulty taking a photograph, and in fact only the steep creek bank prevented me from getting much closer. While I was photographing them the female sidled up to the male and they made a touching picture cuddling up to each other—again very characteristic of Gang-gangs.

They are most interestingly marked cockatoos because of the subtlety of their grey colour and the totally unexpected and contrasting orange-red head of the male. The female is uniformly a mottled grey. While observing these two another female arrived on the same tree and whether out of jealousy or perhaps as a warning, began that distinctive Gang-gang call. Like the Red-rumped Parrots, the Gang-gangs will be looking for nesting sites in tree hollows—usually high above the ground.

Wattle blooms have just about disappeared now and have been replaced by a profusion of smaller flowers. Orchids were noticeable in the barren rocky sections. In the 'worst' type of country—rock with a thin layer of soil—you will often find

Leopard Orchid (*Diuris maculata*).

Wax-lip Orchid (*Glossodia major*).

Pink Fingers (*Caladenia carnea*).

Silver Tea-tree (*Leptospermum multicaule*).

the most prolific displays of wild flowers. Near the track was a small section of exposed granite and gravel which boasted a range of half a dozen colourful wild flowers. There were Wax-lips, purple with white tongues, yellow and dark red Leopard Orchids, Pink-bells, small and delicate pink and white caladenias, white beard-heath, and prostrate tea-trees.

On the topmost point of the ridge above the creek I saw a number of small blue butterflies. One I observed in particular challenged any other male of the same species that approached its vantage point, chased it away and returned to exactly the same position on a dry eucalypt twig. They were Common Dusky Blues. The species feeds on Dodder-laurel, a native plant without leaves and a parasite that twines around other plants, attached by suckers. This congregation of male butterflies on peaks or ridges is known as 'hill-topping' and is common among certain families. It particularly applies to the blues, skippers and swallowtails. The most likely explanation of this phenomenon is that it provides an aggregation of

Black-tongue Caladenia orchid (*Caladenia congesta*).

Phigaloides Skipper butterfly (*Trapezites phigaloides*), a species that hill-tops.

males in a convenient location for females to visit, mate and retreat to lay their eggs on the food plant below.

The evidence to support this theory is that when females are found hill-topping they are unfertilised, and when they mate they invariably retreat. How do the butterflies find their way to the hill top? Is it the silhouette that they see above them? (I've noticed a similarity between the hills which attract the greatest variety of butterflies: they are isolated, rocky, sharply rising, but not necessarily very high.) Do they fly towards the light? Are they lifted by the updraught of warm air as the day advances? It would seem that several factors may be involved because, in my experience, certain species arrive early in the morning (in which case they wouldn't have the advantage of an updraught of air) and others arrive late in the afternoon, particularly on very hot days. One would imagine that these latter take advantage of winds that surge up the rock faces.

The use of hill-top observation in establishing the species in an area is important. A number of butterflies have only ever been recorded on hill tops, and species once thought very rare have proved to be quite common on certain hills.

OCTOBER 22

I decided to go to the high country this morning by way of the transmission line. On the road I took a picture of the main range from a point below Cochrans Gap. In winter and early spring there is an extensive view from here of the snow on the mountains, although it has now disappeared from the lower ranges.

The main range from below Cochrans Gap.

I tried to identify Mount Kosciusko, Australia's highest peak (2228 metres), but this was difficult as it does not stand out, and in fact is only a mere 18 metres higher than nearby Mount Townsend. Other peaks—Mount Twynam (2196 metres) and Carruthers Peak (2145 metres)—look almost as high. In the past there has been some controversy as to which peak Count Strzelecki climbed in 1840, although I believe the evidence very much favours Kosciusko.

The other feature which strikes you about the Kosciusko massif is the bare and rocky outline of the uppermost portion, contrasting with the tree-covered slopes below. The tree-line occurs at about 1800 metres, in comparison to many alpine regions where it may be as high as 3000 metres. This difference is explained by the relatively close proximity of Australia's Alps to the ocean, causing cooler and moister summers at a lower altitude.

Midsummer temperatures are consistently 10°C at the tree-line at various locations around the world. This temperature apparently represents the limit below which there is only sufficient solar energy to provide for respiration and renewal of leaves, and not enough for the growth and support of the branches and roots of a tree.

There was a good wild-flower display along the Kiandra road at around 700 metres. I thought that at this elevation flowers might be a lot later coming into bloom, but because of the cleared area of the transmission line there is much more light getting into the edge of the forest. The flowers were mainly of the pea family—bush peas, Handsome Flat-pea and hovea. There were also a few remaining flowers of coral heath, Small-fruited Hakea and Golden Moths Orchid.

I put my foot unexpectedly on a nest of ants (*Iridomyrmex*). They were in a fighting mood. Their bites were fairly ineffectual but at the same time they gave off a powerful chemical odour not unlike coconut. I've noticed these not unpleasant odours before and I believe they are very attractive to cats! Although most odours emitted by insects are pungent and unpleasant to our nostrils, one must assume that even attractive ones have a defensive role in the insects' biology if they are associated with alarm, as was the case in this instance.

There were a number of bee-flies around the scented flowers. These have a long, thin proboscis and a hovering flight pattern, and in general appearance are similar to bees. They can usually be distinguished from hover-flies because the wings are pigmented.

Flying along the road near a small stream were a swarm of mayflies enjoying their brief nuptial flight. These flies are very familiar to trout fishermen and have two or three long filaments projecting from the abdomen. They have an erratic pattern of flight and I found it difficult to keep track of them in the air.

Because they are so numerous, mayflies are important in the ecology of pond life and in rivers and streams. The nymphal stages of some species feed on the bottom on organic matter while others are carnivorous. They may spend as long as two years in the nymphal stages and having relatively poor defence mechanisms, most fall prey to fish and other aquatic predators—providing them with food.

Their life as adults is extraordinarily short and they may only live for a few hours to take part in the nuptial flight. After mating in the air the female extrudes the egg masses on to the surface of water and they fall to the bottom.

Handsome Flat-pea (*Platylobium formosum*).

Common Beard-heath (*Leucopogon virgatus*) and Pink-bells (*Tetratheca ciliata*).

Alpine Beard-heath (*Leucopogon maccraei*).

I got back to the cottage about 1 o'clock this afternoon and was delighted to be greeted by a flock of Rainbow Birds, swooping and wheeling, trilling all the while. They are always easy to recognise because of their brilliant colour which changes according to the angle of the sun—from gold to green and blue—and from the twin skewer-like black tails, extensions of rear feathers. Their flight is fluent and elegant, a sudden swift dive sweeping into a graceful glide.

Rainbow Birds are insect feeders and they catch them with great precision while flying, and their hunting seems to lead to a certain amount of entertaining aerial competition. They have a migratory pattern from north to south of the continent, and I noticed them when I was at Magnetic Island, Queensland, about six weeks ago, but some have returned south to nest. Others apparently remain in the north all year. Here they favour gullies, formed mainly by erosion, as nesting places. They dig the tunnels in which they lay their eggs into the gravelly soil walls.

Tiger moth (*Asura lydia*).

Ichneumonid wasp, a common visitor to the lights at night.

Tiger moth (*Utetheisa pulchelloides*).

I left the verandah lights on at the cottage last night as the warm humid conditions meant it would be a good night for insects. This morning there were a great number of moths, wasps and other small insects on the walls of the house. Among these were some tiger moths. These ones were aptly named because the forewings are black and yellow, but they had bright red and black hindwings as well. These were not visible because the moths settle with the forewings covering the hindwings. These moths emerge from grass-eating caterpillars which are well known to school children as 'woolly bear' caterpillars.

Tiger moths have evolved a variety of defences. They are unpalatable to predators and during daylight hours 'inform' potential enemies of their undesirable taste by emitting chemical odours. When disturbed they suddenly expose their brightly coloured hindwings. Tiger moths (and some other moth families, too) have sound-detecting organs which pick up the ultrasonic 'echo sounding' noise sent out by bats as they search for their prey, presumably aiding them in taking avoiding action in the event of attack. It has also been shown that some species of tiger moths produce ultrasonic clicks as warning sounds, which cause bats to avoid them in flight. I didn't see any bats last night but I quite often find them sleeping in odd corners inside the cottage.

After lunch I went up to Mocattas Ridge, which leads up to the National Park. This seems to be a meeting place for the dogs that 'worry' our sheep. Three were seen here recently, a tan bitch with two black pups.

I found two flower wasps (Thynninae) mating. The male, which is the only winged partner, holds and flies with the wingless female during copulation. The female hangs head downwards from the male but once copulation is completed, releases itself and burrows into the ground to seek the larvae of a scarab beetle. If successful in this search, the female stings the larvae to paralyse it and lays eggs within it, thereby providing a food source for its own larvae.

The life cycle of some species of flower wasps is closely linked to that of Christmas beetles, which have become serious pests of paddock trees. The adult Christmas beetle attacks the eucalypt foliage but the larvae has a totally different food source—roots and organic matter in the soil. Great numbers are found in the soil beneath improved pastures. With the huge increase in grassland as agriculture has developed, Christmas beetles have multiplied.

The flower wasp also has two food sources. The adult male seeks nectar and honeydew on the eucalypts (and also regurgitates it to the female during copulation) and the wasp maggot feeds on the larvae of Christmas beetles.

The flower wasps in some circumstances are having increasing difficulty exerting control over the Christmas beetles: not only is their egg production relatively low, and they have a single generation in one year, but there is a relatively short limit to the distance the male can transport the female from the food source—on flowers of various trees and bushes—to the breeding site (in one well-studied species, *Hemithynnus hyalinatus*, it is 800 metres). In some instances this means the insect would need to fly to undisturbed areas of vegetation to find plants in flower, which can become an impossible task if too much country has been cleared.

A timber ridge runs along this paddock, part of which forms the boundary with the National Park. An interface of open pasture and forest such as this is very useful if you are observing nature as the edge of the forest is often well endowed with plants. Unfortunately these include weeds, but some native plants also take advantage of the extra light and I found many sundews along the boundary fence.

Tiger moth (*Spilosoma curvata*).

OCTOBER 23

Shearing is in full swing now and is about two-thirds of the way through. Everyone seems to be happy, both with the quality and quantity of wool this year. A dry season will often produce better quality wool and thus far there is no indication of any significant reduction in the size of the clip. The wool looks to be finer, freer from vegetable fault, especially barley grass seed heads, and less affected by rain and damp conditions.

Rounding up Corriedale sheep for shearing.

Green-comb Spider-orchid (*Caladenia dilatata*).

Blue-tongue lizard.

Common Bird-orchid (*Chiloglottis gunnii*).

Crimson Rosella.

This afternoon I walked through an area of bushland on some sheet-eroded soil. A good deal of the top soil had been removed, but low fertility seems to have done no harm to a variety of small native plants, including the Green-comb Spider-orchid. Sometimes the flowers of this orchid are as much as 10 centimetres across, with long slender sepals and petals. It has a prominent fringed tongue with large wine-red clubs, or calli, pivoted so that the slightest tapping of the stem will cause the tongue to vibrate rapidly. Like most native terrestrial orchids they are found in discrete patches and so randomly spread out that one wonders what particular fortuitous set of circumstances are required for their establishment and growth.

Further up the road I came across a blue-tongue lizard tentatively attempting to cross. It was sluggish and only became agitated when I took some close-up photographs. It opened its large mouth and poked out a very blue tongue, hissing and rearing—a fairly effective means of frightening away a bird or animal.

OCTOBER 24

Later today I visited the snow gum country near Tooma Dam. I was struck by its desolate appearance now the snow has melted. The snow has crushed down the low vegetation—tea-tree, bush peas and heath—and as a result old logs, weathered white, have been uncovered. At first glance it looks as though a storm has been through, but all will change very soon as the temperatures rise and the plants begin to grow again and assume their normal upright stance.

Despite the earliness of the season there were a few birds at the 1200-metre level. I saw a number of Pied Currawongs, Grey Currawongs, Crimson Rosellas and kookaburras. At Ogilvies Creek I saw a lone, forlorn-looking Kestrel.

Lower down near Clover Flat I found three species of wattle in bloom: Varnish Wattle, which looks as though some sugary solution has been poured over its leaves; one I have not seen here before, Buffalo Wattle; and Mountain Hickory Wattle.

In an eroded stream bed at Welumba Creek I found a natural kangaroo lick, a section of bared red soil with small excavations in it. Half a dozen kangaroos hopped away as I approached. The soil was more salty than that of the surrounding earth. It is well known that in these hills cattle have always had a craving for salt, and it was a practice of the cattlemen to carry salt with them and to place it at strategic points in their grazing leases.

The weather was again hot and dry which is worrying because October is normally a wet month. With the continuing hot conditions there is every possibility of the pasture 'haying off' early to compound what is likely to be a difficult season anyway.

I found some Pale Sundews in flower. One had caught a large moth which would seem to be beyond its digestive powers. The leaf of the sundew was less than 1 centimetre in diameter and the moth had a wing span of over 2·5 centimetres. There were many other smaller insects on the leaves and a number of mosquitoes and gnats were being broken down so the useful nutrients from their bodies could be utilised by the plant. Moths usually escape the sticky sundew leaves because the surface scales on their wings are readily detached.

Near the interface with the National Park there is a track which dates back to the fire of 1972. This was caused by lightning strikes (about twenty-seven of them) and a large area of the National Park was burnt. The fire was fought on the boundary of this property and the track was cut in to the area by the fire-fighters to enable a back burn. The track marks the boundary of the 1972 fire and one can readily observe what has happened over a period of twelve years. The blackened trunks of the Red Stringybarks and Broad-leaved Peppermints clearly indicate where the back burn was lit and raced in to meet the oncoming bush fire, providing the break which saved a large section of this property.

The recovery of trees and understorey on the burnt side of the track is excellent. This is a characteristic of these so called dry sclerophyll forests. They have evolved over a very long period in an environment of irregular firing and have the capacity to respond rapidly afterwards. The trees send out new growth in the form of epicormic shoots, and the associated shrubs, herbs and grasses seem equally resilient, soon springing back to fill the gaps created by the fire.

Moth (Larentiinae) caught on a leaf of Pale Sundew (*Drosera peltata*).

Violet Kunzea (*Kunzea parvifolia*) among Red Stringybark (*Eucalyptus macrorhyncha*) following fire of several years ago.

Shoots on eucalypt trunk four weeks after fire.

OCTOBER 25

Many moths came in to the verandah lights last night. Sitting under one light was a pair of Emperor Gum Moths, 10 centimetres across the wings. The male was a rich red-brown, the female being rather greyer and perhaps better camouflaged. The male is also easily distinguished from the female by the much larger pectinate antennae—they are quite feathery whereas those of the female, while also plumose, are less obviously so. When I approached the male it became agitated and raised its hindwings to reveal a pair of colourful eyes. You can visualise how, by doing this suddenly and unexpectedly, it could frighten any adversaries. Earlier this year I noticed several of the huge blue Emperor Gum caterpillars with their red and orange spines, on a eucalypt in the garden.

Eye spots on the hindwings of the Emperor Gum Moth (*Antheraea eucalypti*).

Male Emperor Gum Moth showing the large feathery antennae.

These moths communicate by pheremones, often produced in very tiny amounts. The female releases them from organs on her abdomen and the male with antennae covered with sensory cells is able to detect the minute amount of pheremone molecules circulating in the air. Females are capable of attracting males from surprising distances.

Interest in pheremones has increased in recent years because of their potential for controlling pest insects, particularly moths. Several which are quite specific have been identified chemically, and some have been manufactured on a commercial scale and used successfully to monitor fluctuations in the population of pest moth species, in order to improve the timing of sprays. The use of synthetic pheremones to directly control pests, either by trapping of males to reduce the population, or by spreading through an orchard to confuse them and so prevent breeding, has not yet succeeded commercially.

This afternoon I went to Mittamatite, a range which rises rather unexpectedly between the Murray and Corryong Creek. Although outside my immediate area of interest it provides a good vantage point and I thought it useful to visit, if only to gain a broader perspective.

I went to Embreys Lookout, a point giving commanding views of Corryong, Mount Kosciusko and the Dargals. The elevation of the mountain is just under 1000 metres and Embreys Lookout a little lower. The rocky terrain now supports an abundance of spring flowers. The attractive Nodding Blue-lily was conspicuous with its yellow anthers. The plant is well named with the royal blue flowers drooping down from the stem. Also prominent was the Variable Groundsel—a bright yellow daisy—and the two flowers provided a pleasing touch of colour on the granite outcrop.

I found a *Grevillea jephcottii*, an unusual grevillea with dense green and red flower clusters, found only on Mittamatite, nearby Pine Mountain and Mount Burrowa. It was named in honour of Edwin and Sydney Jephcott who collected many plants last century for Baron von Mueller in the vicinity of Pine Mountain.

I came across some of the more unusual insect galls. In this case the male and female galls, which protruded from the leaves of a small eucalypt, were very different in appearance. The male gall was perhaps 1 centimetre high and was a good imitation of a minute golf tee, and the female much larger and almost the size of an acorn. The gall-forming insect in each case would be contained within the structure. The response of the plant to the chemical secreted by the male is obviously different from its response to that of the female. It is fascinating to think the plant can create these structures and reproduce them more or less exactly time after time in response to the irritation set up by different sexes of one insect.

Nodding Blue-lily (*Stypandra glauca*) and Variable Groundsel (*Senecio lautus*).

Male and female galls of *Apiomorpha conica* on eucalypt leaf.

Caterpillars of the Emperor Gum Moth.

November 3

Heavy falls of rain last week have improved the farming prospects markedly. Although the pastures are by no means good for spring at least the rains have secured some sort of hay crop.

It was much hotter today than on my previous visit to the farm. There were plenty of birds on the Murray flats and I saw several large White-necked Herons. There has been a dramatic increase in the number of Wood Ducks on the front dam of the home paddock. There are now about fifty birds, compared to the twenty we had earlier in the season. It is difficult to identify the young from the older birds as they all look much the same now. Despite the fact that these birds are on the main thoroughfare into the property and are passed by traffic all day long they remain very wary, perhaps because they are a target during the duck season. As you pass they fly off together in a tight formation to another dam about 400 metres below, then return until the next passing car sets them off again. Wood Ducks feed on pastures and the improved pasture around the dam appears to provide most of the food they want. I imagine they may be nesting down on the Tooma flats in the large River Red Gums.

Wood Ducks.

White-necked Heron.

Violet Kunzea (*Kunzea parvifolia*).

Common Fringe-myrtle (*Calytrix tetragona*).

Violet Kunzea and Common Fringe-myrtle where forest joins farmland at Welumba Creek.

Case moth (Pyschidae) and meat ant (*Iridomyrmex purpureus*) on Golden Everlasting (*Helichrysum bracteatum*).

At Welumba Creek the purple kunzea and fringe-myrtle, largely white but sometimes pink, were out in profusion. There were a great number of insects and birds in attendance, particularly around the fringe-myrtle. It gives off a powerful honey odour, but there seems to be nothing specific about its pollinators—it attracts everything, and today I saw beetles, native and introduced bees, moths, butterflies and even some plant bugs on the flowers.

On the Golden Everlastings alongside the track there were quite a few insects, particularly meat ants. On one flower I noticed a case moth caterpillar feeding alongside several ants, each intent on their individual tasks—the caterpillar eating the bracts and the ants looking for nectar.

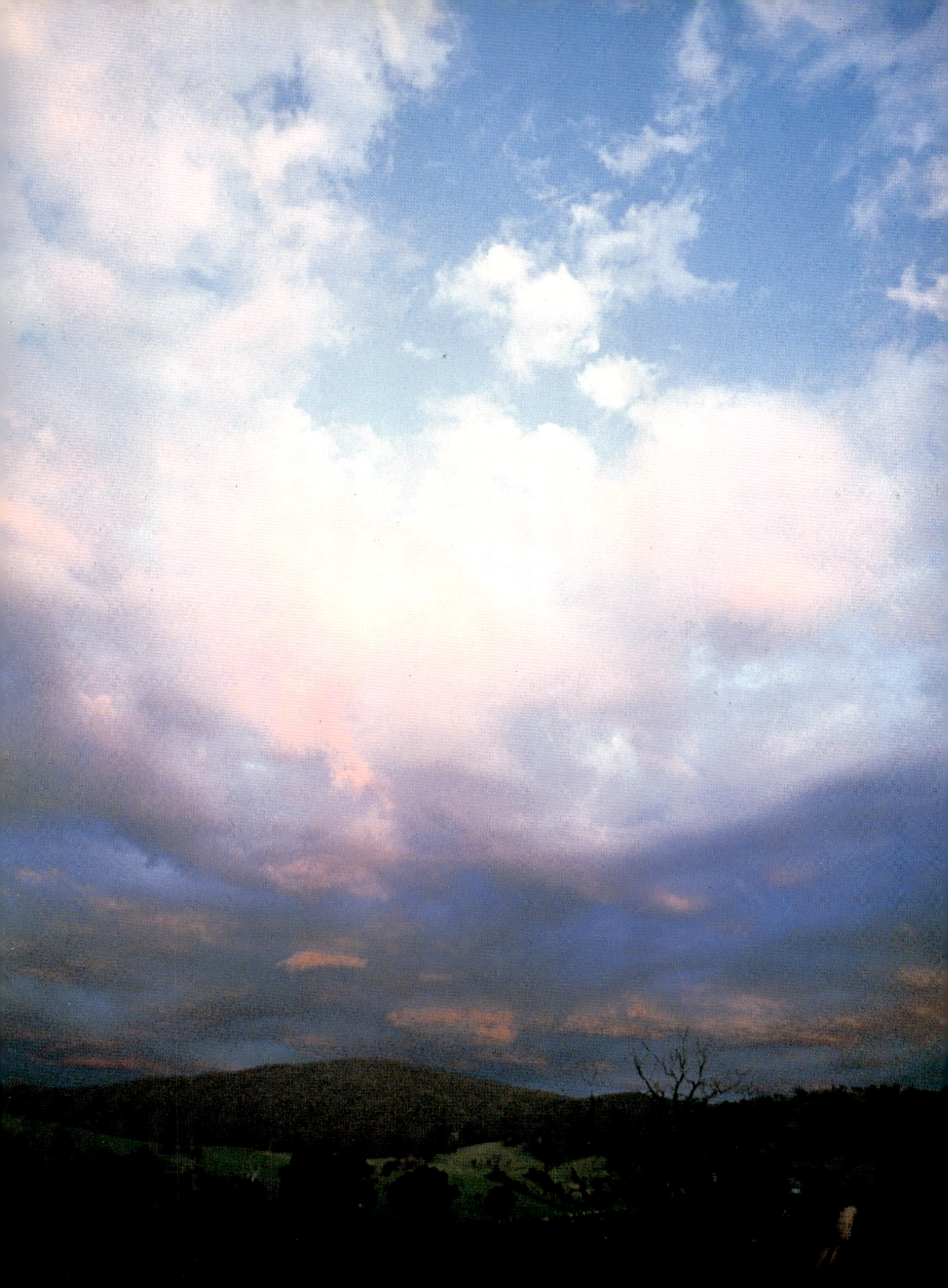

November 4

Today I photographed a number of flowers, including the prickly tea-tree (Manuka), with its pure white flowers and sharply pointed leaves. These particular specimens were more like climbing plants, covering rocks just a few centimetres above the ground. Among the many other flowers on display were bluebells, rice-flowers and fringe-lilies, the latter the first I have seen this year. This species, the Twining Fringe-lily, is lilac, with a distinct fringe around the petals and sepals. In full sunlight its appearance is enhanced by the satin texture of the flower parts.

While I was looking at the wild flowers I disturbed a couple of Forest Bronzewings which set off with a whirring noise. These birds, the first I've seen up here, are readily identified in flight by the bronze patches on the wings. They are essentially seed feeders, concentrating on the seeds of wattles and other legumes. I guess they have been attracted by the wattles which, now that they have ceased flowering, will be beginning to produce seeds.

In the afternoon we took the children swimming in Taylors Creek which runs through the farm. On the way we saw a crow apparently chasing a starling. At first this seemed a fairly innocuous incident and it was not altogether clear who was chasing who, but it soon became evident that the much larger crow was the aggressor. Suddenly it darted downwards, grabbed the starling and took off. It alighted on a fence post, with its victim held in its claws, and was joined—at a distance—by a number of other starlings, seemingly very upset by the attack. By the time we arrived the crow had set off again. It is the first time I have seen a crow actually catch a bird in the air.

Common Fringe-lily (*Thysanotus tuberosus*).

November 5

OPPOSITE PAGE:
Spring sunset from the cottage.

Today we added one more orchid to our list—the Common Beard-orchid, easily recognised because of the hairy nature of the tongue. It was a solitary specimen on the most barren piece of ground you could imagine, not one other plant within several metres of it. These orchids are pollinated by a species of hairy flower wasp.

I caught sight of a goanna on a small stump within a metre of where I was standing. It was so well camouflaged that I almost missed it. When they are frightened goannas often run up a tree and this one had picked a very short one indeed. Having run out of tree it 'froze' on the top. It was a colourful goanna with a lot of green on the body and a yellow throat. When I drove away, it hurriedly retreated and presumably found a much higher refuge.

Goanna on tree stump.

I had arranged to take the children camping at Welumba Creek and we left about 5 o'clock with the weather looking very threatening. After about half an hour of rain it cleared and we were able to set up camp.

During the evening I looked at several wattles which I've had under observation for some time. I was aware that the Moonlight Jewel butterfly caterpillars were living on one of these trees and I was able to find a caterpillar making its way up to feed on the foliage. During its slow progress it was accompanied by about eight ants which swarmed over it as though giving it encouragement and direction.

The female of the Moonlight Jewel lays her eggs from early December through January on the trunk of the wattle, in close proximity to a nest of these ants, in my experience usually on the easterly side of the tree. Possibly the butterfly locates the ants by their characteristic odour. As soon as the caterpillars emerge from the eggs (which are sculptured rather like sea-urchins) the ants take them down into the nest. They protect them and accompany them on warm nights up to the top of the tree to feed. During the winter the caterpillars hibernate and begin to feed again as the weather warms up in August and September. They pupate as a rule in early November and hatch from late November through to about Christmas time.

Eggs of the Moonlight Jewel butterfly.

Moonlight Jewel butterfly (*Hypochrysops delicia delos*).

Larva of Moonlight Jewel being 'guided' up the trunk of a wattle by ants (*Crematogaster sp.*).

November 6

We returned to Melbourne today. Interestingly the heavy rain we experienced at Welumba Creek last night was very local indeed. The cottage, in the neighbouring valley of Taylors Creek and only 5 kilometres away, did not have a drop.

November 26

All the hill country pasture and much of the flats are now quite brown and dried off. There are patches of green shoots where hay has been cut and in the soaks along the river. November is the month which often shows the most radical seasonal changes here. The mean average temperature rises from about 15° at the beginning of November to over 20° by December. Unless it is very cool or there is exceptional late spring rain, which is about one chance in ten, the light green of late October changes to light brown in a matter of three weeks, or sometimes a few days.

The paucity of bales that have been cut in each paddock, now ready to be moved into stacks or enclosures, bears witness to the poor spring. Although the quality of the hay is not always good, it is an important commodity, given the uncertain autumn rainfall. If the autumn break arrives late, a good supply of hay can make the difference in seeing out the long cold winter.

As I drove from Melbourne early this morning and crossed the Murray at Towong Bridge I saw White Ibis flying in typical V formation and wondered how they so easily maintain this flight pattern. The swamps and billabongs along the Murray have been much reduced during November, causing ibis to congregate so they're easier to see than a few weeks ago.

The number of Wood Ducks has dropped to around twenty again. They tend to disperse early in the spring to mate, but will probably come back to this favoured camping site in autumn. In our experience some will stay right through winter.

Towong Bridge.

As if by magic the spectacular wild-flower display at Welumba Creek has now disappeared. There is barely a single plant of kunzea or fringe-myrtle in flower although their place has, to some extent, been taken by tea-tree and dogwood. There are still a great number of yellow everlastings along the track.

One insect was in great numbers — a species of tiger moth. They were feeding quite openly on the yellow everlastings, displaying their orange and purple wings. I also found a cup moth caterpillar, particularly large and bright green, yellow and purple. They are often called Chinese junks and are armed with two pairs of retractable organs at each end. Unfortunately I became so engrossed with its potential for a photograph that I stood in a dog trap. This is a possibility I've always been concerned about, but it proved to be a fairly minor incident. I had a pair of gumboots on and while it hurt slightly, I was more startled than injured.

Dog traps are set in a cunning manner, buried under the ground with soil and leaves placed over the top of the jaws. The wild dog becomes curious about the disturbed ground around the trap which is set on their chosen pathways. Many people regard trapping as cruel, and I guess it is, but it is the only practical way of reducing wild dogs along the boundary of the National Park where shooting and poisoning are not permitted.

Alongside a Slender Tea-tree, on the bank of the creek, was a thicket of Small-leaf Bramble. It has a superficial resemblance to a blackberry, but the fruit is like a raspberry. The flowers are very attractive to insects, but what struck me was that

Seed head of Variegated Thistle (*Silybum marianum*).

Robber fly (Asilidae) with its moth prey (Lymantriidae).

Cup moth caterpillar (*Doratifera sp.*).

the insects lured to the bramble were quite dissimilar to those frequenting the tea-tree immediately alongside it. Different kinds of flowers attract different kinds of pollinating insects. There were no beetles or tiger moths on the bramble, but there were several species of native bees, introduced bees and four species of butterflies.

I located an eagle's nest on the Lighthouse Mountain which I've been looking for for some time. The main peak of the mountain is about 730 metres and the name Lighthouse is said to stem from the early days of settlement when the fire from a surveyors' camp, located on the western side of the mountain, appeared as a suspended light high above the Murray to those on the other side of the river.

The nest is an impressive structure and belongs to a pair of Wedge-tailed Eagles. I wouldn't like to estimate the weight of wood these birds have managed to accumulate on this slender tree, but the nest would be 2 metres in width and perhaps 1 metre deep. You can just see into it from below the peak, but it is not in use at present. Wedge-tails line the inside of their nests with fresh leaves but the leaves in this nest are now old and brown. It's a shame we didn't discover it three or four months ago because it could well have been in active use. There have been Wedge-tailed Eagles constantly flying around this peak, and I noticed young ones some months back.

Sheep camp on top of the lower Lighthouse peak, which was completely cleared in the early days. As a result the soil nitrogen levels are high, and there is a profusion of weeds, mainly Slender Thistles. Today these thistles were being patrolled by fierce-looking robber flies, predators with hairy legs and long proboscises, which they thrust into their insect victims, often while on the wing.

I rescued a butterfly which had been grasped by a robber fly. The butterfly was quite inert, although it obviously had recently been very much alive. Robber flies paralyse their victims then suck the juices from their bodies. They inject enzymes to make them easier to digest. Later I saw other robber flies and, in all cases, the insect that they'd caught was quite lifeless yet apparently fresh.

NOVEMBER 27

This morning I noticed one Willy Wagtail that was unusually interested in the front verandah and I found that it was nesting there. The nest is a small hemispherical structure, made from grass and spiders' webs and with an almost polished outer surface. Inside were three small mottled eggs. I was able to photograph them, but the female became quite aggressive. Eventually it flew away, but I noticed it later sitting on the nest and didn't disturb it again.

Young Willy Wagtails in the nest.

Willy Wagtail on the vine near its nest.

There are dozens of Welcome Swallows flying around the cottage. I don't know why there should be so many—I can find only one nest near the roof of the machinery shed. I think it is in use although its height prevents closer inspection.

I found my first Christmas beetle today. Normally Christmas beetles are common at this time of year and you will see them flying around the top of the red gums. The adult beetle and the larvae live on two entirely different foods: the larvae on organic matter in the soil, the adult on eucalypt leaves. Some idea of the abundance of these insects comes from the CSIRO work at Armidale in the New England district where as many as 24 000 larvae per hectare have been found under improved pasture.

I think the reason there are so few Christmas beetles about this year may be because it is unusually dry. In a very dry season beetles have great difficulty in emerging from the ground. It has also been observed that they are rare two years after a summer drought. In a drought the hard soil prevents the establishment of the larvae in the soil. The female lays her eggs but the young larvae are unable to penetrate the surface. The beetles normally have a two-year life cycle and the effect of the adverse conditions shows up two years later in a reduced adult population.

Christmas beetles may make conditions more favourable for other defoliating insects later on. They are usually the first to attack paddock trees and the subsequent regrowth is attractive to other insects, particularly the chrysomelid beetles.

I vaguely realised that the front verandah at the cottage had become much lighter since my last visit and discovered that the leaves of the ornamental vine had been completely removed by voracious vine moth caterpillars. I'm not quite sure what this will do to the vine, but I don't expect many of the colourful leaves next autumn. Whether it can re-shoot at this stage of the year would seem doubtful; a lot will depend on how much rain we have.

Over at Welumba Creek it was clear that the lyrebirds' behaviour has changed with the season. They seldom run through their repertoire during the spring months. Lyrebirds gain maximum impact by mimicking bird calls out of season—during the winter breeding period, when the birds they imitate are silent—presumably so their song will be recognised as that of a lyrebird. Although I could not hear them there was no doubt they were in the area because of the prolific scratchings.

I found two interesting insects. The first was a moth (Zygaenidae), closely allied to the burnet moths of Europe. A brilliant metallic blue, this moth was sipping nectar from a bright yellow everlasting and it moved only slightly as it extracted the nectar.

Christmas beetle (*Anoplognathus sp.*).

Moth (*Pollanisus sp.*) on Golden Everlasting (*Helichrysum bracteatum*).

The best find was an Apple Hanging Moth. I've been looking for these for some time and I found a male, newly emerged and resplendent in apple green with silver markings, hanging in characteristic pose by his forelegs, next to the pupal shell protruding from the stem of a small dogwood. It will hopefully mate this evening because, like other members of the swift moth family, this species seldom lives much more than a night, although two years pass from the time the egg is laid until the appearance of the adult.

It has always intrigued me as to how the Apple Hanging Moth, a member of a family of moths that normally spray their eggs on the ground, manages to lay eggs on small branches of trees; then how the larvae burrow through very hard bark and establish themselves inside the stem. I recently found that the life history has been untangled by work done in New Zealand. A similar species there has been found to have three separate larval phases. In fact, Apple Hanging Moths, like other swift moths, do spray their eggs on the ground. The tiny larvae find a hiding place beneath stones and small logs, living on organic matter and fungi in their first two instars. A 'transfer' stage larva of different appearance then develops and seeks out a site on a branch, and gains entry by way of a tunnel it cuts. A third phase larva, different again in appearance, lives perhaps a year and a half in the branch before emerging as an adult moth.

Male Apple Hanging Moth (*Aenetus ligniviren*).

Moth (*Euloxia sp.*) attracted to light at 1300 metres.

Forest Phebalium (*Phebalium squamulosum alpinum*).

Rain drops on wattle pods.

Beetles (*Diphucephala sp.*) on Leafy Bossiaea (*Bossiaea foliosa*).

NOVEMBER 28

I camped overnight near the Tooma Dam. When I woke about 4.45 a.m. the glow of the sun was just appearing over Jagungal, beneath some ominous storm clouds. Fairly steady rain had fallen throughout the night and though it was still wet I had a look around outside the hut. Despite the poor light at this hour it was obvious that there had been a great change from three weeks ago, with numerous plants now in flower, principally lemon-yellow phebalium and orange bossiaea. They were flowering on the verge of the snow gum copses, on shallow soils overlying the granite. After fire these two shrubs have a great capacity to colonise bare areas between the snowgrass tussocks.

November 29

The cattle and sheep all look in good condition at this late stage of spring. The pastures, however, are thin, and it will be necessary to feed hay to the bulls and probably to some of the autumn calving cows after Christmas.

Beyond the Tooma Dam this morning I stopped to look at an Alpine Ash stand. The understorey is made up essentially of one plant, Hop Bitter-pea, which has become dominant as a result of past light fires. It is somewhat inconspicuous when not in bloom, but now there are masses of orange flowers on bushes about 2 metres high. The plant, as the name suggests, has a very bitter taste, and the chemical responsible has been extracted and used in the past for medical purposes. It was also used as a substitute for hops. One characteristic of the plant which makes it recognisable among bitter pea species is the triangular seed pod.

I decided to camp out overnight on the Lighthouse. On the way up I noticed several Kestrels and there was one flying above the peak. They are a superb bird to watch in flight, russet on top and almost cream underneath with streamlined swept-back wings. This Kestrel was gliding up against powerful wind currents from the west, and was so able to regulate its flight that at times it hovered without motion. On other occasions it would rise over a hundred metres, seemingly without moving its wings, and then slowly descend again to hover in exactly the one position. Suddenly it would drop like a stone almost to the ground, to pause again with a tremulous yet rapid wing beat a few inches above the surface. I'm not sure what particular animal or reptile it was after—possibly young rabbits or more likely mice, native rodents or perhaps lizards. However, in the time that I watched it was not successful in its hunting.

Preying mantis (Mantidae) lying in wait.

Mantispa (Mantispidae) on grass-tree.

Mayfly (Ephemeroptera) on grass-tree leaf.

Immature male phasmatid (*Didymuria violescens*), a species that on occasions reaches plague proportions, causing damage to eucalypt forests.

After setting up camp and having a meal I took the torch and went out hopefully to look for native animals. There were no signs of any, although the night was quite warm and seemingly suitable. There were, however, many insects on the grass-trees and one which took my interest was a mantispa. It has an uncanny resemblance to a preying mantis and yet it is actually a lacewing. It has delicate filamentous wings in which the veins are clearly delineated. Its resemblance to the mantis is not some form of mimicry and one assumes that it has evolved this way to give it the advantage that such highly developed forelegs provide for an insect that lurks in wait for its prey.

November 30

Huntsman spider (Sparassidae).

This morning I had a look at an area of red gums (*E. blakelyi*). A number were laden with creamy-white blossoms, alongside others that were obviously not going to flower this year. What triggers flowering at one particular time and why aren't neighbouring plants all similarly stimulated?

I pulled some bark off the red gums to see whether there were any insects or spiders and was greeted by a number of large huntsmen spiders. They are quite frightening in appearance, but relatively harmless. One fell on me and ran up my neck. My experience is that they don't bite, although if you encouraged them they probably could give you some sort of a nip.

I disturbed a large flock of starlings on the track to the cottage. I continue to wonder at the complete harmony of their movement. How is it that a hundred birds can regulate their flight at such speed so their position relative to one another is maintained, each wheeling and banking movement carried out in perfect unison, enabling them all to alight on a tree almost in the one instant? They are birds that can raise the ire of both farmer and bird lover alike, and they are thought to compete for nesting sites with many native birds. It is believed that the smell of the grass after it has been used to line nesting hollows in trees may well repel parrots and other native birds from using these sites. Starlings came originally from the northern hemisphere and were introduced in 1850. Unfortunately, unlike some other introductions, they've extended beyond the cities and now occupy quite considerable areas of farmland, although they do not appear to have invaded the wetter forest areas.

Flock of starlings.

As I was leaving for Melbourne I drove out of the property through the large round bales lying in the paddocks. We use them a lot nowadays because they are much more economical both to cut and to feed out. Standing about 1·5 metres high, they look strange dotted over the river flats and the rolling hills. But they weather well and last at least six months, though it is generally accepted that it is better to bring them in to an enclosure and to feed them out, rather than have the cattle and sheep 'attacking' them in the paddock in late summer or early autumn.

Newly made round bales.

DECEMBER 16

This morning at dawn along the Tooma River wisps of mist followed the outline of the red gums along the banks. They rose like lazy spirals of smoke from early morning camp fires.

As I came down over the gap separating the Murray and the Tooma, dawn light was picking up the brome grass flower heads along the side of the road. Prominent too were the russet-red docks extending in places into the paddocks. The tallest weed was Twiggy Mullein, like some thin yellow-flowering hollyhock. The long spikes were covered with green buds and just a few yellow flowers with black centres opening at the base. It is sometimes a nuisance here on the rocky hills, particularly where they have been heavily grazed. Looking through an article tonight I see that the seeds of a closely related species are remarkably durable, and some kept in a bottle for eighty years subsequently germinated.

From Towong Bridge south towards Cochrans Gap and further to the east over the Dargals and the Grey Mare Range, the play of hazy light and shade picked out the worn contours of the ancient eroded hills. The flats have started to dry out. There is a network of green grass and weeds which fills the water courses, billabongs and ponds, making a sharp and pleasing contrast with the russet of the dock and the light brown of the cut pasture.

There are now far too many rabbits about, particularly young ones. A few have myxomatosis, but unless the disease hits very hard in the next few weeks they will be a problem in the new year.

Twiggy Mullein (*Verbascum virgatum*).

Dragon fly (*Diphlebia sp.*).

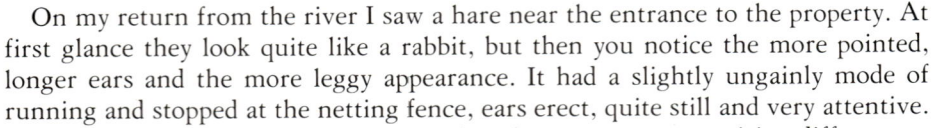

On my return from the river I saw a hare near the entrance to the property. At first glance they look quite like a rabbit, but then you notice the more pointed, longer ears and the more leggy appearance. It had a slightly ungainly mode of running and stopped at the netting fence, ears erect, quite still and very attentive.

Hares of course are related to rabbits but there are some surprising differences. The hare does not live in a burrow, but uses an open hollow or depression in the ground called a form. The hare in contrast to the rabbit is born with fur and with its eyes open. It is not affected by myxomatosis or poison baits.

Willy Wagtails were busy at an early hour this morning. I looked at one as it sat on the neck of a cow, close to its head, watching intently to see whether any insect would be uncovered as the cow grazed. Later on I noticed these same birds on the fence outside the kitchen window, their tails wagging continually, making sudden full turns and then swinging back again — always moving. Every now and then one would fly off the fence and attempt to catch an insect in mid air.

The most pervasive sign of summer is the smell of dry grass. It's one of those marvellous smells that evoke memories of past summers and the special quality of the Australian countryside.

Later this morning I watched dragon flies (Amphipterygidae) patrolling one small section of Welumba Creek, the brilliant turquoise males chasing other males away and returning to settle on the same rock. They rely entirely on colour to determine the sex of their fellow dragon flies, and in contrast to other insects, pheremones are thought to play little part in their mating behaviour. I saw a male and a female flying in tandem, each grasped to the other — a preliminary to copulation. I unsuccessfully searched for the nymphal stages of these particular dragon flies — strange squat creatures living on the rocks in a fast running stream. They lie completely submerged and have three swollen tracheal gills at the end of the abdomen, enabling them to breathe while under water and at the same time providing means of 'jet' propulsion when they need to make a fast retreat.

In a neighbouring territory just off the stream another species, black and yellow with green eyes, seldom ventured out over the water, but patrolled the banks within about 10 metres of the edge.

There was considerable bird life around this morning. I noticed two Sulphur-crested Cockatoos flying at a great height — I'd estimate at between 150 and 300 metres. Their habits are quite variable and you see them in almost all locations at different altitudes and in numbers varying from single specimens to literally hundreds in one flock.

The amount of earth turned over shows lyrebirds are still about. I am a little concerned for their future here because of the prevalence of wild dogs and cats. Nevertheless the lyrebirds have obviously learned to live with these two predators over the years, although they have not done so in Sherbrooke Forest near Melbourne, where their numbers have dropped alarmingly in recent years.

Flowering brome grass (*Bromus sp.*) along the roadside.

After lunch I climbed the northern face of the Lighthouse to see what changes have taken place during the spell of warmer weather. I was surprised how little regeneration there has been of Silver Wattles since the drought of two years ago. I could only find two live trees among the dozens of dead ones. I think it will be a long time before they get back to their original numbers on these shallow stony soils. Root suckering may have failed and rabbits have possibly eaten those seedlings which have germinated from seed, although there must still be an abundance of durable seeds in the ground.

Flock of Sulphur-crested Cockatoos.

Coming over a rocky ridge I encountered a group of three kangaroos: one male, a much smaller female and a joey that looked too big to get back into the pouch. Fortunately they did not see me and I was able to duck down behind some rocks, take a quick light meter reading and set my camera. As I rose stealthily to take the photograph I noticed how alert the female was. She didn't see me as I hid behind a rock but she looked around, apparently sensing that something was not in order. The male was less sensitive and the joey quite disinterested. I was able to take a series of photographs before the female saw me. Since I remained perfectly still they didn't move off immediately, but looked at me with great curiosity which soon merged into fear and they hopped away. There's always been a small family of Eastern Grey Kangaroos living on the Lighthouse and I frequently used to see an old man kangaroo—about 2 metres high—which had become separated from the mob. The much larger mob near the boundary of the National Park seems to be distinct from this group.

Sulphur-crested Cockatoos.

There were few flies for this time of year—you are usually surrounded by a horde of small bush flies, not unlike house flies, which take great joy in sucking whatever available moisture there is on your skin. They are native insects that have thrived since European settlement some two hundred years ago. They can survive in a wide range of temperatures, from 12°C to 35°C, but at 30°C and above their activity is greatly increased. The larval cycles are much quicker, and the flies breed more frequently.

119

Bush flies are very dependent on animals for moisture, carbohydrate and protein. They cluster around your eyes because sweat and tears are rich in materials they can utilise. They are also dung feeders. They seek out fresh dung because the eggs will not produce larvae unless the conditions are humid. After reaching maturity the larvae move into the ground and pupate a centimetre or so below the surface.

Temperature more than any other factor apparently controls the population of bush flies. In southern Australia temperatures during the cooler months are not sufficiently high to support a permanent population and flies migrate south each year from the drier and warmer inland areas on the irregular northerly winds. They then tend to build up during late spring and again during autumn. Numbers are controlled by the type of dung in which they breed. If the animals are feeding on new sprouting pasture and their dung is relatively soft and nutritious, the resulting bush flies are larger. They lay more eggs so the population increases with the prospect of more food (i.e. dung) ahead of them. If, on the other hand, the dung is fibrous, the larvae produced are small, the flies correspondingly small and they lay fewer eggs. Thus the population declines if the prospects are poor. Animals and birds can on occasions be predators of bush flies because there are so many of them in the environment, but generally they are singularly free of enemies, either predators or parasites.

I found a large ant lion on one of the lights outside the cottage this evening. These lacewings somewhat resemble damsel flies or dragon flies with flimsy-looking large wings and quite small bodies. This species has swept-back wings, marked in white and dark brown.

Ant lions are so named because of the voracious nymphal stage. Although this particular species does not build traps, its larva is a predator armed with a powerful set of jaws and a fat, dumpy body. Some species construct a conical pit, 2 to 3 centimetres deep, to trap insects. The sand from which the pit is made is at the angle of repose so that the insect, once trapped inside, finds it impossible to escape. It keeps trying to climb to the top, but rolls back to the bottom and eventually a pair of pincer-like jaws emerge and grab the hapless insect, usually an ant, and drag it below the surface. It seems strange that such a diabolical creature could give rise to a delicate, inoffensive-looking adult.

Ant lion (*Periclystus circuiter*).

DECEMBER 17

This morning I spent some time among Mountain Swamp Gums at the foot of the Dargals. While there are not a great many plants in flower now, Smooth Tea-tree was much in evidence along the creeks. Few of the white flowers are attracting insects, their nectar supply virtually exhausted.

On the remaining attractive flowers the insects were dominated by the large flies (Tachinidae). These are like giant blowflies, but much more hairy and often brightly coloured. The larvae are largely parasitic on beetles, sometimes on the larvae of moths.

I attempted to photograph restless damsel flies at a small pond in the forest. I subsequently realised that the pond had been artificially formed, possibly during the grazing years, and was really an old dam that had partly filled up. On the way back I had a momentary encounter with a Red-bellied Black Snake, which was about 2 metres long and thus fully grown. It looked sinister as the sun caught the lustrous black upper surface and gave a glimpse of the bright red underside. I was impressed by the effortless liquid movement of the snake. As soon as it sensed my presence, about 3 metres away, it reared up, momentarily flattened its head and then spun around and glided back to the edge of the pond, a typical habitat. They are partial to frogs and there would certainly be quite a few here. Red-bellied Black Snakes, in my experience, tend to be most active in the late afternoon although I've also seen them at other times of the day.

I haven't seen many snakes this year. Normally you would expect to encounter quite a few during the spring months when they're more sluggish and fail to sense your vibrations as you walk about. Black snakes and the brown snakes, certainly at lower elevations, seem to occupy the same territories, but the browns are faster, more aggressive and much more dangerous than the blacks, which have considerably less toxic venom.

Copperhead snake on the road near Clover Flat.

Brown snake on the Snakey Plains Track.

I stayed out late tonight near Bradneys Gap picnic area and found the mosquitoes aggressive and persistent during the evening. The creek would provide an ideal breeding ground for the larvae. Many species of mosquitoes are able to breed in small areas of water accumulating in clefts in trees, rock holes or hollows in the ground. I thought it a good thing that only the females need blood, to ensure maturation of their eggs, and thus the males don't worry you. Not all species of mosquitoes attack man and many are quite specific to native animals and even to reptiles and amphibians.

The kookaburras are noisy tonight. It's hardly surprising since kookaburras become addicted to the bits and pieces available at camping or picnic areas. Although they're traditionally said to seek out reptiles, lizards and snakes, they in fact have a much wider dietary interest. They'll eat insects, even small birds, and are very partial to scraps.

DECEMBER 18

Purple Violet (*Viola betonicifolia*).

I woke before dawn hoping for a good day in the mountains, but the clouds were blowing up and it was obviously going to rain. Knowing the weather can be very fickle at this time of year, I thought it best to have a look at the high points early and then retreat if and when the weather turned.

Near the turn off to the Snakey Plains, the flowers were out in abundance. It's surprising what an extra 1000 metres of height will do. The season here is many weeks behind the flowering at Welumba Creek. Certainly some plants are different species, but the spring at this elevation does not occur before December.

There were many daisies, principally *Brachycome decipiens* and *B. spathulata*, with a mauve ring of petals (ray florets) surrounding the yellow centre. There were also Pink-bells, Gorse Bitter-pea, Alpine Oxylobium, Billy Buttons, some species of buttercups and white caladenia orchids, to name but a few.

Tufted Daisy (*Brachycome scapigera*) and Granite Buttercups (*Ranunculus graniticola*) near Tooma Dam.

Native bee (*Lasioglossum sp.*) asleep on unopened flower of *Brachycome spathulata*.

Ivy Goodenia (*Goodenia hederacea* var. *alpestris*).

Clematis (*Clematis aristata*).

I found a number of bees which had apparently slept out overnight on the daisies. There they were, comatose in the yellow centre of the flower, surrounded by the folded purple florets which hadn't yet responded to the early morning sunshine.

I looked for signs of an alpine form of hanging moth, *Aenetus paradiseus montanus*, a species which normally flies in October and November. It's an important source of food for black cockatoos and this was clearly demonstrated in my search. Although I could locate the position in the stems of the snow gums of many first year caterpillars, all second year mature ones had been attacked by the cockatoos which had unerringly found the caterpillars or pupae and extracted them. It made me wonder how the moth could survive such an onslaught, but they've both been around for many thousands of years and somehow have struck an equilibrium. Most of these moths (Hepialidae) accept huge losses in their immature stages, and of the thousands of eggs that are laid by a single female only a few will survive. Usually the losses take place at the egg stage or in the early development of the larvae. However, in this species control by predation of the more mature larvae by the black cockatoos must be a prime method of population control. Considerable damage is done to the snow gum saplings by the black cockatoos as they bite into the branches so that in places it looks as though they have been attacked with a small tomahawk.

It was interesting to note the prolific regeneration of snow gums in this open woodland which was old grazing country, not far from the stock route across the site of the Tooma Dam and up to Round Mountain. The grass and herbs seem to have reasserted themselves effectively but there was not much shrub development. Even where the bossiaea, phebalium and oxylobium are now established, they will probably enter a declining phase in a few years, opening at the crown and allowing the snowgrass to grow through and eventually take over.

The rain caught up with me and I was forced to retreat and I stopped at Clover Flat at an elevation of just over 900 metres. There were cascades of white clematis (*C. aristata*) over trees and shrubs, which is common throughout the higher rainfall areas. This climber is a member of the buttercup family (Ranunculaceae) and is similar to exotic clematises grown in gardens. Unlike many vines it causes no real trouble to supporting plants because it is light, using a tree for support yet not smothering it.

The creek at Clover Flat is bordered by banks of tall dense Mountain Tea-tree. Beneath the tea-tree and next to the fast running stream are a few Mountain Peppers. These belong to the magnolia family and have red stems, dark green leaves and inconspicuous brown flowers. You can easily identify them by the extremely hot taste of the leaves.

DECEMBER 19

I set off to Round Mountain after lunch today, hoping to compare the flora and fauna at 1500 metres with what I saw on my trip yesterday, which took in elevations of 1000, 1300 and 1500 metres.

On the way I stopped again at the Snakey Plains turn off and noticed that many of the older Mountain Gums and snow gums had pronounced fire scars on the south-eastern faces of the butts. This would at first glance suggest a fire driven by wind from the south-east, but in a bush fire, flame vortices driven by wind tend to converge on the lee side of the butt. The fact that the trees were on a slope would encourage a hot fire. This fits the most likely pattern you would expect here, with a fire burning from the west or north-west. The fire scars are also on the uphill side of the tree, where bark and leaves tend to accumulate, which further encourages damage on that side.

I came across some jumper ants' nests. Jumper ants, with yellow jaws and black abdomen, have the capability of jumping and delivering a particularly vicious sting. The nests were covered either with small cubic pieces of charcoal from burnt snow gums or pieces of red punk from rotted tree trunks. This contrasts with the situation lower down in the swamp gum forests, where the same ant covers its nest with similar sized pieces of white gravel. Apparently it is not the material that is important, but the size of the particles.

I arrived at Round Mountain at about 7 p.m. and set up my tent on the ridge where I camped with the children last May. A strong, almost gale force wind blew up and I decided to repair to the car for the night.

DECEMBER 20

This morning the car is covered with ice and hail. It's blown like fury all night long and the tent is only just standing. It is 6 a.m. and there is not much point in staying any longer. Conditions are now so bad that it's quite impossible to do anything worth while.

I arrived back around 8.30. It's blowing a gale here, too, but although we're only about 25 kilometres away from Round Mountain, as the crow flies, there's no sign of rain. As I look back from the cottage, the Dargals and mountains beyond are completely covered in cloud.

On the cottage verandah the vine that was completely denuded by vine moth larvae last month has made a startling recovery. The leaves are already in excess of 5 centimetres in just three weeks. Vines are extraordinarily vigorous and resilient plants, with roots going down great depths and able to tap resources of water and nutrients from a vast volume of soil.

Despite the weak pastures the stock are in good condition. They often look their best during the drier months and hit peak condition in January and February. If it is a very good summer they continue to improve right up to the autumn break. In poorer seasons such as this, they start to tail off as the feed runs out.

I counted fifteen magpies in the front paddock on an area of 25 hectares—apparently a high density for magpies. I believe there is usually a minimum of 4 hectares for a breeding pair, and it's apparently uncommon to find a group of more than ten together. I gather there are seldom more than three adults and two young ones permanently in a single territory. However, during the period after nesting, while the young ones are being 'trained', magpies may accumulate in larger numbers, and I guess the situation will change as they mature and go out on their own.

The young Willy Wagtails in their nest on the front verandah have grown dramatically. In just five days they've changed in appearance and now have grey feathers streaked with white and handsome brown heads. They raised their heads

Female Yellow Spot Jewel butterfly (*Hypochrysops byzos hecalius*).

Green cockchafer beetle (*Diphucephala sp.*).

Eastern Ringed Xenica butterfly (*Geitoneura acantha acantha*).

Male Common Brown butterfly (*Heteronympha merope merope*) sunning itself in first rays of dawn.

on extended necks with mouths wide agape, but the female returned, sounded its alarm call, and they retracted their heads in an instant.

I noticed several small metallic green cockchafer beetles, some on crinkled leaves of pomaderris. This is also the food plant of the Yellow Spot Jewel butterfly (*Hypochrysops byzos hecalius*). The caterpillars are difficult to see, being well camouflaged as they rest on the pale hairy undersurface of the leaf. The best way to locate them is by the small round holes they eat in the leaves, unlike the more irregular indentations made by beetles and other species of caterpillars.

DECEMBER 21

I went to Mittamatite at dawn this morning to take some sunrise shots. The point I chose, a gateway 450 metres above and immediately west of the junction of the Tooma and Murray rivers, provided some good views. Groups of awakening Common Brown butterflies settled in the first rays of sunshine on an embankment opposite the gateway, the light creating a brilliant orange iridescence on their brown and black wings. Within twenty minutes their bodies had warmed sufficiently for them to start flying.

There were no female Common Browns about. It has been shown that after mating the females go into a resting phase over the summer months—a mechanism presumably to ensure the later survival of their caterpillars. During the period of rest the eggs develop. The females begin to fly in February, and lay their eggs as late as March or April, hopefully by which time the grass on which the caterpillars feed will be green and able to support the next brood.

The Golden Everlastings at Welumba Creek are now well past their prime. There are still hundreds of flowers along the track, but they are showing signs of wear and tear; each fruit (achene) with its bright yellow tuft of hairs, or pappis, is starting to disengage from the flower to be swept away by wind.

Later in the afternoon I climbed back to the area that I'd looked at earlier in the week on Lighthouse Mountain. I had seen an interesting orchid in bud and wanted to photograph it with its flowers fully out. It was a Hyacinth Orchid. Today several of the lower buds had burst open to reveal vivid red flowers mottled with darker shades of red. This orchid is leafless, with a reddish-brown stem. It derives nutriment from its relationship with the mycorrhizae of fungal associates on its roots. Unlike the mycorrhizae associated with trees, which form a sheath around

Sunrise from Mittamatite, above the junction of the Tooma and Murray rivers.

the lateral roots and only penetrate between the cortical cells, the mycorrhizae of orchids actually invade the cortical cells. The Hyacinth Orchid, with little or no photosynthetic tissue, must depend on the mycorrhizae for the provision of sugars. It is not known with certainty how this transfer of glucose is achieved.

Since the Hyacinth Orchid is always found near eucalypts, it was once thought that in some way it was parasitic on their roots. This hypothesis has subsequently been refuted, but it does seem that the appearance of the mycorrhizae may be connected with the close proximity to eucalypts.

How arid and generally bare this part of the country now looks. It is a typical northern face, emphasising the difference in climate between the north and south aspect of these hills. Earlier in the morning on the Mittamatite massif, certainly at a higher elevation but on a southern face, the conditions were much milder, the grasses more prolific and not so dry. The Common Brown butterflies on the Lighthouse Mountain are very worn, whereas those on Mittamatite were fresh and undamaged. In seasonal terms the difference could be as much as three weeks or more, because the northern face is hotter and more exposed than the corresponding area on the other side of the valley.

Close-up detail of Hyacinth Orchid (*Dipodium punctatum*).

JANUARY 14

We have had very little rain now since October and following the dry winter there is some fear of yet another drought, although this is an 'assured' rainfall area. In the last twenty years we have suffered two of the most serious droughts since the district was settled. In 1967 rainfall was little over 350 millimetres and in 1983 only reached 330 millimetres. On the other hand, we have had several years in excess of 1000 millimetres.

A study of local rainfall records indicates that the drought of 1914–15 was perhaps the most devastating. Looking at the records which were kept by Mr Arnold Playle in 1915, one gets some idea of the severity of the drought. Of course it has to be said that in those days there were far fewer watering points and roads, and rabbits were a much more serious problem than they are today. Nevertheless, his graphic description of finding dead kookaburras under trees and magpies so 'tame' they actually came inside the house seeking water, gives some indication of the extreme dryness. In April 1915 he wrote:

> Driest autumn known here by white man. All grass and food gone, cattle all falling away rapidly. Barley and other green feed sown for winter use is still lying in dry dust and there is no moisture to germinate it. Several laughing jackasses and other birds were found dead under the trees, no insects about, only ants. The three large creeks are running again, the Cudgewa Creek stops for 5 or 6 miles, the Thowgla Creek for 3 or 4 miles, and the Corryong Creek for 7 or 8 miles, and finally reached the Murray on the 17th of this month.

Early summer morning.

Eventually, after almost a year, the rains came in May 1915 and he wrote:

> There were good useful rains and light frosts. The green shade was coming over the ground like a single coat of paint where it was not too bare. Stock horses, cattle and sheep are dying in hundreds. The ploughing and sowing are being done wherever possible. I found kookaburras, honeyeaters and other birds dead under the trees. All the streams are now running, water is in the dams but none has got in the wells.

OPPOSITE PAGE:
Fence detail.

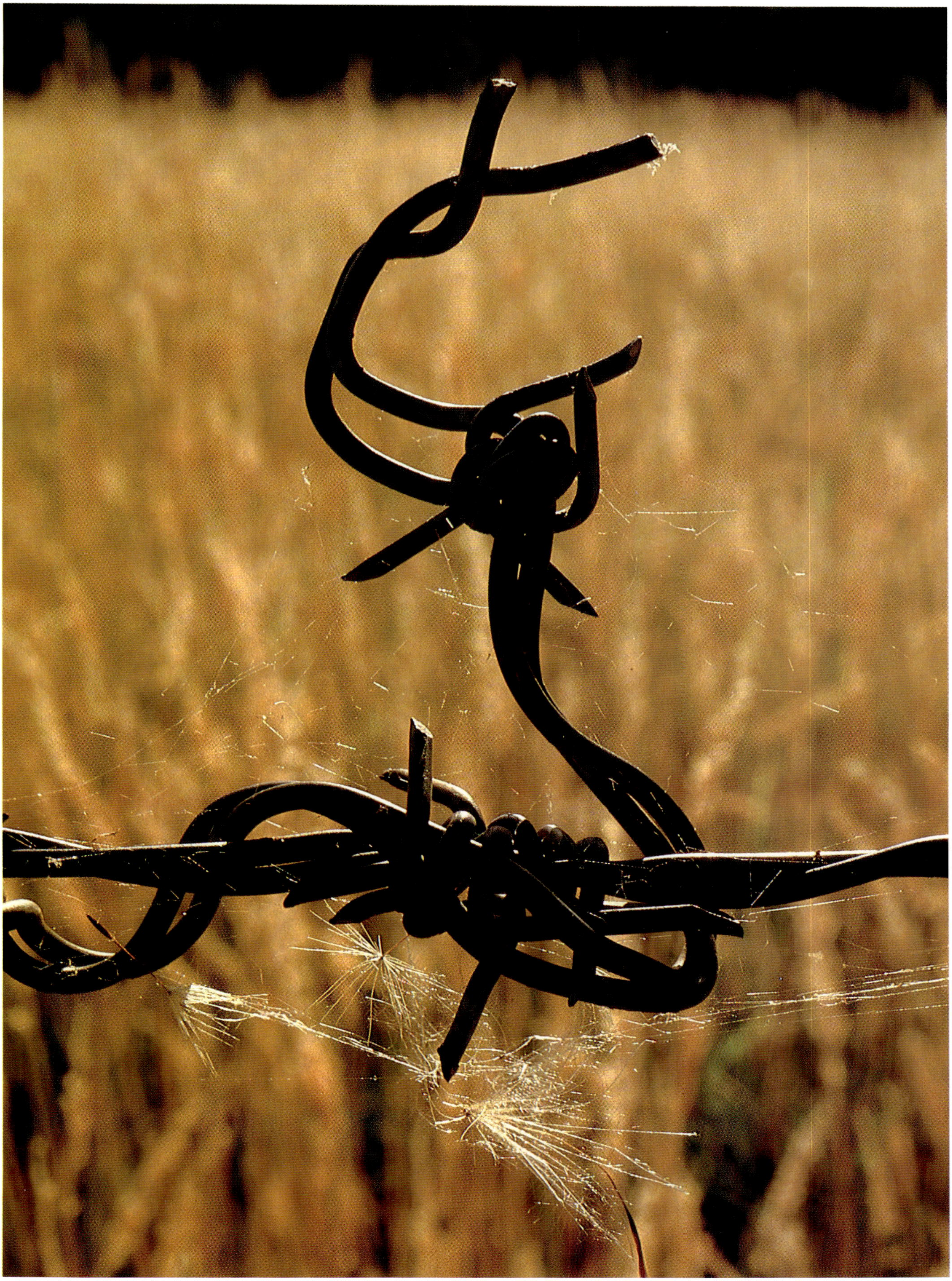

Today promised to be very hot so I sought the cooler temperatures of the high country. The view towards the head waters of the Murray, across Towong Hill station, shows the sown grasses are much whiter and hungrier-looking than last month. The Kangaroo Grass is in flower and brown patches of it stand out on the dry hills. A few pockets of green still remain along the water courses of the Murray and the Tooma.

Kangaroo Grass (*Themeda australis*).

Setting out to muster cattle.

There's a good variety of flowers under the snow gums at the 1200-metre level, the most prominent being Alpine Oxylobium, a vivid orange flower, and prostrate kunzea with small yellow flowers creeping over the ground. I also noticed one Golden Moths Orchid and a few buttercups still in bloom. The flowers that dominated last month, like Forest Phebalium and Leafy Bossiaea, are on the wane. Near Pierces Hut there were a number of blue daisies (*Brachycome*) and the first Grass Trigger-plants I've seen this summer.

A few snow gums are blooming, the sparse flowers small and white, some with the operculum or cap just coming away and revealing the white stamens.

I camped at Round Mountain—on the same ridge as last month, when I was forced to leave early after a blizzard during the night. Today I think the temperature must have been above 30°C.

Robber flies were out in big numbers. One I photographed had captured a bug and pierced it with its large evil-looking proboscis and carried it easily in the air. Flower spiders were everywhere on the daisies. They are strongly territorial and stay patiently for hours on the one flower. I found a butterfly, looking quite life-like although uncharacteristically hanging from the underside of a flower. Closer examination revealed that it was suspended in the jaws of a camouflaged flower spider, the butterfly quite comatose and no doubt soon to be eaten.

Over the top of the ridge at Round Mountain, in sight of the huge solitary peak Jagungal, there is a moss bed and I found groups of Candle Heath (*Richea continentis*) in discrete patches, growing directly out of the mat of sphagnum, the white flowers attracting a great number of insects. As the name suggests they stand erect with large, column-like inflorescences and the leaves are prickly and rigid. Candle Heath is one of a group of heaths best represented in Tasmania, there being only one species on the mainland.

Later in the day I climbed Round Mountain. I walked across about 1 kilometre of open plain and sphagnum bog. I was pleased to find that the sphagnum has re-established itself well with large patches covering several hectares. I reached the base of the mountain and worked my way through scrub which became increasingly thick, until the bossiaea and Alpine Orites were 1 to 2 metres high. Progress was

Bark detail, snow gum (*Eucalyptus pauciflora*).

Nymphal shell of a cicada (*Diemeniana euronotiana*) shares a daisy with a fly and a looper caterpillar.

Flower spider (*Diaea sp.*) with paralysed prey, an Orichora Brown butterfly (*Oreixenica orichora orichora*). The flower is Variable Groundsel (*Senecio lautus*).

Snow gum (*Eucalyptus pauciflora*).

slow and arduous and became even harder when I ran into dense clumps of Mountain Pepper. I was glad to reach the small narrow fringe of grass which surrounds the immediate summit. The last 60 metres or so over rock was a lot easier than bashing through scrub. Round Mountain is capped by basalt and in places the surface is bare rock, almost moonlike in its shape and structure.

I arrived back at the cottage at about 4.30 p.m. to catch the tail-end of a heatwave. I think it's been over 40 degrees.

A huge flock of Galahs had gathered just opposite the cottage, much bigger than I've seen here before. They seem to be attracted to the Slender Thistles. The large flocks like this one are made up of immature birds, not yet territorial and covering large areas.

At about 10 o'clock this evening we had a dry thunderstorm, preceded by a violent wind which blew the pictures down in the living room. I raced to close all doors and windows but not before the lamp had been overturned and papers blown in all directions. Shortly after there was the rumbling of thunder and some vivid flashes of lightning. A few minutes later I could see a glow on the horizon to the east and another to the north-east. Undoubtedly these were small fires started by lightning strikes. Later there was a slight smell of bush-fire smoke in the air, but in half an hour the fires had been put out by the local fire brigade. They're well organised and keep in constant contact during periods of high fire danger.

JANUARY 15

There is a big fire burning near Jingellic 30 kilometres down the river and another 15 kilometres to the south-east in the Bringenbrong State Forest. Although it is much cooler this morning, with a lot of light cloud, there is an almost continual rumble of thunder. These summer thunderstorms are often associated with much noise and a great deal of lightning, but bring little rain. There are always a few each year in the Upper Murray, and they are accompanied by bush-fire outbreaks. These days few summer fires in this area are started by people as the dangers are well understood, but lightning strikes remain a real threat.

Red gum (*Eucalyptus blakelyi*) after the storm.

The damage from last night's wind storm was severe. Almost all red gums on the property were affected in some way. One or two old trees were snapped off at the butt, and most lost a few major limbs. On the ridges there were large numbers of branches down, and yet in other places trees had miraculously survived. On the ridges it could be that exposure to wind makes the trees vulnerable, or perhaps they lack moisture from growing in shallow soil and hence are more brittle.

The other most common tree in the paddocks is Apple Box and they were only mildly damaged, though they don't appear any stouter or more robust than the red gums. I noticed the leaves on the fallen red gum branches were quite wilted even at 6 o'clock in the morning, eight hours after the limbs fell. The leaves of the Apple Box were only slightly wilted, which could suggest they hold more moisture than the red gums. It has always been a tenet of bush lore that you don't sleep under a red gum in summer, and there have been several fatal accidents when large limbs have fallen on camp sites and cars.

The farm animals and bird life seemed quite unaffected by the stormy weather and smoky atmosphere. Again there were many Galahs, particularly immature birds with little pink on them. These young Galahs form roving groups looking for whatever feed they can find. Larger groups are more likely to find patches of grain or thistle heads than smaller bands or individuals. I understand that the basic unit is a breeding pair and these don't roam very far. The immature birds, on the other hand, are likely to cover a much wider territory. Whether their increased numbers this year are the result of the good season following the drought or whether they are vagrants from another area is difficult to say.

Galahs feeding on thistle seeds.

The Willy Wagtail fledglings have left their nest, which now looks a little bedraggled. The mature birds continue to fly near the nest as though they still have some interest in it. Are they going to breed again? This does happen and they may have several broods in one year. At one stage today the Willy Wagtails became quite agitated and I wondered whether this was because I was sitting near their old nest or whether it was due to the Kestrel perched in a nearby tree. Perhaps the immature wagtails, which I have not seen for some time, have fallen prey to a hawk or a Kestrel.

Examining some eucalypt foliage today to see what has been eating it, I discovered a superbly camouflaged green looper caterpillar (Geometridae) 'standing' by means of its rear claspers, rigidly straight at an angle from the stem. Its tapering outline was exactly like a shoot. Earlier in the year I found a caterpillar of the same family on a brown dead twig, again exactly matching the colour. In this instance, the blunt head of the caterpillar echoed the end of the broken stem next to it. A brilliantly striped red, black and white caterpillar (Agarastidae) near by fed openly on the leaves of mistletoe, apparently 'confident' that its warning colours would keep birds away.

Looper caterpillar (Geometridae).

Caterpillar (*Comocrus behri*).

Caterpillar (Lasiocampidae).

Looper caterpillar (Geometridae).

There's nothing quite like a bush-fire sunset. The sun takes on a deep crimson, almost purplish hue when viewed through heavy banks of bush-fire smoke, and I waited for an hour or more to photograph the big red disc of the sun as it sank below the outline of the Koetong Plateau.

A large menacing-looking preying mantis 'parked' close to a verandah light tonight, undoubtedly hoping to feast on a few of the moths that flew in. One of its huge raptorial front legs was held more advanced than the other, rather like a boxer 'shaping up'. As I moved my hand towards the mantis it rocked and swayed, presumably to frighten me. It is interesting to consider why mantises should adopt a camouflaged appearance. Is it because they are trying to hide themselves from predators, or is it to conceal themselves from their insect prey?

A Short-horned Grasshopper came to the light. It too was well camouflaged, with long straw-coloured legs and abdomen to match the dry grass. It is impressive how these insects adjust their appearance to exactly match the seasonal colour of the vegetation.

Setting sun through bush-fire smoke.

JANUARY 16

Still the thunder continues to rumble. At times it is so continual as to be like distant artillery. But there's no real cloud bank and no immediate prospect of rain.

Skink lizard minus its tail.

The fire is still burning in the Bringenbrong State Forest, causing some worry. The fire brigade was out last night trying to contain it with a back burn. The forest adjoins the Kosciusko National Park and the prevailing westerly could push it into the park. Farmers on the perimeter have established breaks which will probably stand up to a resurgence of the fire.

The Jingellic fire also continues to burn and is providing most of the pall of smoke now covering the district, but we understand it remains under reasonable control and the only threat would be from a recurrence of the hot north winds.

I noticed a skink lizard sitting patiently on a rock in the creek. They remain motionless seemingly for hours waiting for the odd insect that comes near enough for them to pounce. When they do attack they move extremely quickly.

On a track on the property I came upon a colourful group of jewel bugs, metallic blue and green with red spots. They are leaf-sucking bugs which emit a strong odour if alarmed. The meat ants—in hundreds on the track—took no notice of them, and even one walking across a meat ants' mound was shunned by the swarms of ants. They did not, however, shun me as I photographed this interesting little scene.

On the grassy outlet of the dam I disturbed a Blue-winged Shoveler, which flew off, beating its wings against the water. Shortly after I noticed at my feet another one, an immature, concealed in the reeds. If I had not been looking closely I would have missed it. As soon as it realised its camouflage had been discovered it took off in characteristically noisy fashion.

I found a dead sheep surrounded by a number of crows on my way back to the cottage. It appeared to have been attacked by a dog, and I found out tonight that this was true. In many sheep attacked by dogs up here, evidence of the wounds can be missed unless you observe the carcass closely.

Some huge thunder heads developed over the mountains in the late afternoon—great white billowing clouds, but the forecast indicates that there is no relief in sight and we urgently need rain to put out the fires.

Bush-fire sunset reflections in the dam.

Thunder heads over the Snowy Mountains.

JANUARY 17

This morning I had a closer look at the storm-damaged trees. It is evident that many of the breakages occur in the red gums at points previously injured, either by insects, by past fires, or where dry rot has occurred.

The stock on the property all look in great shape. The sheep have pelted up well since shearing in late October. They seem able to thrive on the dry grass, clover and medic burr. In warm conditions the energy requirement of stock is much less and providing they have ample water they do well.

Even in this dry spell there's still plenty of water coming over the waterfall at Taylors Creek. We always notice any lag in the stream flow. Late in a dry summer the streams may still flow quite strongly, fed by water stored up from the heavy spring and late winter rains.

This afternoon there were a few birds behind the cottage. I could hear treecreepers with their distinctive and repeated piping note. There were Eastern Rosellas, the usual magpies, Galahs and Sulphur-crested Cockatoos and two or three Whistling Eagles. The latter were a graceful sight circling the eastern flank of the Lighthouse in great wide sweeps. They're essentially carrion and insect feeders and I guess the prospect of dead rabbits would be attracting them. Their flight is slower than some other birds of prey, and you can pick up the characteristic kite tail.

Before sunrise, Tooma River.

This afternoon I brought back a large stem of flowering Apple Box. It was in its prime and gave off a strong smell of honey. I stuck the spray in a vase of water on the kitchen sink. Later I was met by many small black ants which had obviously left the flowers and found their way down over the lip of the vase. I had not seen them at the time I picked the stem; presumably they were inside the flowers feeding on the nectar.

This made me think about the role of ants in pollination. Although you frequently see masses of them on flowers it has been generally thought that they are not important as pollinators. I used to think this was because their inability to fly would make them much less efficient than their cousins, the wasps and the bees: they wouldn't be very effective crawling down a tree or a branch and then crawling up another one to reach the next flower. However, from my recent reading it seems that the situation is considerably more complicated. It has been shown that some of the chemicals secreted by ants, and which tend to cover their bodies, inhibit fungal development and are thus of value in their nests. However, the chemicals would seem to do the same thing to pollen development; in other words, they prevent the development of the pollen tube. Presumably for this reason ants have not shared in the mutually rewarding coevolution that has occurred between flowering plants and wasps, bees and butterflies.

The air is still smoke-laden, mainly from the Jingellic fire, though the Bringenbrong State Forest continues to burn. I believe that there were forty-five lightning strikes in the district in the last forty-eight hours, which gives some idea of the extent of the problem.

As I make these notes at 10.45 this evening the wind has turned and is now coming from the south. The weather is quite cool and hopefully will stay this way.

JANUARY 18

It is much cooler this morning, and the sky is clearer, although there is still a good deal of smoke.

I took a short walk up the eastern flank of the Lighthouse where there were some Bursarias in flower. Among the insects on the plentiful white flowers were Fiddler Beetles, and flower chafers with a brilliantly enamelled sheen—yellow with black and dark brown markings. I watched one flower chafer, a little more than a centimetre in diameter, feeding. Splay-legged, it worked its way through the blossoms, its palps in constant motion, sensing out the nectar.

Several meat ant mounds showed the unmistakeable signs of an Echidna's presence in the last few days—large surface excavations which exposed the many galleries beneath. Echidnas, which are not uncommon here, are closely allied to the Platypus, both being classed as monotremes. The spur on the lower leg of the male Echidna is not poisonous, unlike that of the Platypus. This Echidna had presumably been catching ants with its long sticky tongue. How it can digest these and termites, too, is quite remarkable. I searched in vain for a glimpse of the animal, but in my experience they're seldom seen, even where they have been recently working. They hide readily under stones or logs and if no shelter is at hand they can disappear straight down into the ground at amazing speed.

I noticed a mud dauber wasp (Sphecidae), a black and yellow creature nervously moving over rocks, looking for spiders. There are several mud dauber nests around the cottage. They paralyse spiders, carry them to their nests and encase them inside a cell of dried mud. The mud dauber is easily identified by its extraordinarily thin waist or pedicel which separates the enlarged abdomen from the thorax.

Later I found another wasp, a small metallic green species, inside the fly-wire screen of the kitchen window. I picked it up and it immediately rolled itself into a ball. This was a cuckoo wasp (Chrysididae) and a parasite of the mud dauber and would have emerged from larvae in one of the mud dauber nests on the verandah. Presumably the cuckoo wasp rolls itself up for protection against the jaws of the much bigger mud dauber wasp. The smaller insect is well protected, almost armour plated, and the jaws of the large mud dauber are not likely to affect it. The relationship of the cuckoo wasp and the mud dauber provides an example of secondary parasitism, a parasite living on a parasite.

Flower chafer beetle (*Polystigma punctata*).

Echidna.

I visited the Yellow Boy, a clearing in the mountains behind the property that was grazed last century. It is said that near here Baron von Mueller, in January 1874, found a new species of plant, *Bertya findlayi*, which he named after his host James Findlay. Findlay at that stage was at Towong Hill and the Baron made a trip to the Dargals with him.

Yellow Boy is a subalpine meadow hidden away and dominated by Mount Jagumba, which rises to the north-east. The Yellow Boy Creek runs along the edge of the meadow and is well clothed in Mountain Swamp Gums, Blackwoods and tea-trees. Both kangaroos and wallabies abound here.

Later on this afternoon I climbed the Lighthouse and spent some time trying to take a photo of the Yellow-winged Locusts, which rejoice in the Latin name of *Gastrimargus musicus*. Presumably the name comes from the curious clicking noise they make while flying. It's a hard species to approach as it seems to have a sixth sense about your presence.

Yellow-winged Locust (*Gastrimargus musicus*).

The most striking thing about this grasshopper is the colour of the hindwings—bright yellow with black markings. These, of course, are classic warning colours, but whether they are used to frighten or confuse predators (probably birds) is debatable. A bird seeing this obvious black and yellow insect in flight will concentrate on the same colours after it has settled. But when the grasshopper alights it is almost perfectly concealed by the leaflike forewings and the bird remains perplexed as to where it has gone.

I remember Malcolm Burr describing this in his book *The Insect Legion*. He had some chickens running 'wild' in a camp in Africa and when they followed the local locusts with bright red hindwings, never once could they find them. They would set out with eager anticipation but as soon as the red was replaced by the camouflage beige, the chickens were completely nonplussed.

Fruit of the Small-leaf Bramble or Wild Raspberry (*Rubus parvifolius*) growing along the Yellow Boy Creek.

The Yellow-winged Locust is not generally regarded as a pest of pastures here, but in some years it is a nuisance. This year it has been confined to native pastures; in other seasons it has moved on to improved pastures reducing the amount of feed available to stock. We had a minor plague in 1972, a dry year when we needed every bit of feed we could find. This grasshopper is often associated with the small wingless grasshopper, which is even more damaging and takes a fancy to the garden as well as the paddocks.

A swarm of meat ants were running up the trunk of a red gum, within a few metres of a large nest with a number of 'paths' leading out into the grassland.

Meat ants (*Iridomyrmex purpureus*) at entrance hole in nest.

About 8 metres from the mound I discovered one ant 'manfully' attempting to move a large grasshopper's head, several times its own size, to the nest. To test how accurately they can track their spoil, I removed the grasshopper head from the ant and placed it about a metre away. The ant was unable to find the head but it was soon found by another ant that proceeded with it towards the nest. This ant made good progress despite the size of its burden, carrying it over large tussocks and small branches. Eventually it, too, became confused. At this point another approached, seized the head and took off again with much more purpose. However, within 2 metres it also became disorientated and moved around in circles, still retaining the head. As before, other ants approached and one seized the head. This ant had obviously pricked up the pheremone trail better than the others and made unerringly towards the nest. When it reached the mound it was immediately advanced on by a number of ants who proceeded to tug at the head. It shook all these off, passed three entrances and made towards its own entrance hole where it deposited the head.

Two aspects impressed me. First, the ants are clearly dependent on scents or pheremones to enable them to follow a trail. In broken, rough country like this and at a distance from the nest the trail can easily be lost. Secondly, the ants act for the benefit of the colony and not directly for themselves. The food the worker ants bring back to the nest will be used to feed the larvae. Nevertheless, each ant feels impelled to contribute to the food gathering process and they 'fight' for the honour of being the one to bring in the 'prize'.

Foraging worker ants, after locating food, walk back to the nest with their abdomens touching the ground at intervals, laying a pheremone trail. This enables other workers to follow the trail to the source of the food. The trail, however, is not long lasting. If it was other workers would continue to act instinctively and search after the food source had been taken away. The ephemeral nature of the pheremones ensures that no such wasteful activity continues after the food source has been gathered up and taken to the nest.

JANUARY 19

A few fires are still burning and the sun came up through the usual smoky haze. The authorities claim the Bringenbrong State Forest fire is under control, although they admit that it has jumped the Khancoban–Kiandra road, which is now closed. However, it was a cool morning with little wind.

There were quite a few Superb Blue Wrens about. You always see them in groups, though they also live as separate breeding pairs. This morning they were hopping about, two males in full plumage and eight dull-coloured birds which could either be females or immatures to my less than expert eye.

It would appear that the blue wren relies heavily on living as a group to improve the survival rate of the new brood and as a means of recovering from periods of adversity, such as a drought. The immature birds assist in the feeding of the young, enabling the female to concentrate on starting the next brood.

Apparently it is only the dominant male bird, usually a survivor of four or more winters, that tends to retain its blue plumage throughout the whole year. The immature males lose their blue and go into what is called eclipse plumage closely resembling the dull-coloured females.

When the younger males change into their new colours, and become sexually responsive, breeding is already under way. The rivalry that one would normally expect to occur between the males is muted by the fact that there are now young birds to be raised in the new brood. The young female birds are eventually forced out to seek their own territories, and thus are subjected to greater dangers than the immature males, which stay with the original group.

I decided to have a last look at the high country before going back to Melbourne. The alpine flowers are now out in profusion, with banks of Derwent Speedwell

Bush fire, Bringenbrong State Forest.

Derwent Speedwell (*Veronica derwentiana*).

Derwent Speedwell growing among snow gums above Cabramurra.

along the roadside and under the snow gums, the white and lilac tinged flowers sparkling against the green understorey of herbs and grasses. Various species of daisies dominated the snowgrass openings between the gums.

On a snow gum leaf I noticed one hairy caterpillar (Anthelidae) carrying a number of small red mites. These mites are often found with certain moth caterpillars, attached to their surface.

Snow gum leaves, especially those on the seedling trees, were attracting a variety of leaf-eating insects. These were mainly beetles of the family Chrysomelidae. There were also a lot of plant bugs (Coreidae), which have a particularly pungent smell when disturbed, vaguely like some apple juice concentrates. They were spectacularly marked—red with long legs, a black outline to their abdomen and a yellow underside.

I was struck by the slow recovery of vegetation along the roadside where the road-making operations of several years ago are still obvious. The only plant that appeared to be making any headway was a branching prostrate *Kunzea ericifolia* with bright yellow flowers.

Anthelid moth caterpillar (*Chenuala heliaspis*).

Grass Trigger-plant (*Stylidium graminifolium*), the lower flower with trigger released.

There were several large groups of particularly richly coloured, deep magenta trigger-plants among the snow gums. The flowering head, usually about 7 to 8 centimetres long, contains flowers at different stages of development, the older flowers at the base, younger flowers in the middle and buds at the apex. I picked up a piece of grass and inserted the stem in the centre of one of the mature flowers and immediately, seemingly from nowhere, came the trigger, snapping across the open flower. There are five petals, the fifth one being the labellum, minute and hidden below the poised trigger. After about fifteen minutes the trigger came back to its resting position but it was not possible to get it to perform again. It takes about thirty minutes for the turgor to return to the stem.

Beetle (*Anoplognathus sp.*) on snow gum leaf (*Eucalyptus pauciflora*).

Chrysomelid beetle (*Poropsis augusta*) on snow gum leaf (*Eucalyptus pauciflora*).

Damsel fly (unidentified) resting by an alpine creek near Cabramurra.

The trigger is a column made up of the stalks of the anthers and the style. Looking at a young flower closely the pollen bearing anthers are prominent and conceal the style. In older flowers, however, the style has taken the place of the withered anthers. When a species of bee, the principal pollinating vector in south-eastern Australia, visits the flower, it sets off the trigger. In the case of a young flower the bee will be sprayed with pollen, will take it away and cross-pollinate another flower. Later the same flower will in effect become female when the pollen is exhausted: when the trigger falls across another visiting bee the flower will collect pollen from it and thus itself be cross-pollinated.

It was very hazy and cool at this altitude today, above the bush-fire smoke. On the previous days the smoke must have come over the top of the range, because I found blackened leaves dropped from the Jingellic fire. I saw these falling from the sky in the still evening air at the cottage a few days ago. They are scorched rather than burnt leaves. Apparently they break off from the tree as the fire approaches and are caught in the powerful updraught, and propelled thousands of feet up into the atmosphere, finally coming to earth many kilometres from the fire.

A similar mechanism may apply to burning bark, and fire can be transported well ahead of the front. This process is known as spotting. It is a characteristic of eucalypt forest fires and makes their control extremely difficult under acute conditions of high temperature, strong wind and low humidity.

Spotting with fibrous bark, such as is found on stringybarks and peppermints, does not usually involve distances beyond 3 kilometres. However, with Alpine Ash the large dangling pieces of bark, often 5 to 10 metres in length, can be transported up to 25 kilometres ahead of the blaze.

On the way back I stopped at some burnt snow gums above the town of Cabramurra. This section of snow gum woodland was obviously burnt in a very hot fire some years ago. There is not much regrowth and the dead timber is silvery white. The mass of burnt driftwood-like spars contrasts with the rest of the landscape. Because of the cold conditions and the natural hardness and durability of the timber it is preserved down to the smallest branches.

Dangling strip of bark—which can be carried by wind in a bush fire and cause spotting.

Burnt snow gums (*Eucalyptus pauciflora*) and regrowth, Cabramurra.

JANUARY 20

The sky is now much clearer and most of the smoke has gone.

The Welcome Swallows were back in force this morning, many of them on the television aerial and the eaves of the cottage. They were joined by some Wood Swallows. One Welcome Swallow flew up to within a metre of me in an attempt to catch a small fluttering moth. I was sitting on the verandah at the time and it hovered quite near my shoulder in a pose not unlike that of a humming bird, except that you could easily see its wings beating. It failed to catch the moth but demonstrated wonderful manoeuvrability and balance.

Turning over some old sheets of corrugated iron I found the first Red-back Spider I have seen here for several years. Fifteen years ago they were very common and could be found under almost any piece of dry wood picked up in the paddocks.

I took some photographs of an area on the property where three stages of development meet—improved pasture, native pasture and the original stringybark forest. The Kangaroo Grass in the native pasture is now a rich burnished red-brown. It's interesting to see the stability of this pasture; it is resistant to weeds and to re-invasion by other native plants, although only a fence separates it from the improved and fertilised pasture sown down in the late 1960s.

The news from the fires is much better and it seems that the danger period is past. There is concern about the forecast for stronger winds, but providing we can get through tomorrow the fire-fighters and the farmers should have some respite.

Red-back Spider (*Latrodectus mactans*).

Sunrise on the Murray River at Towong Bridge.

It is almost a year since I began this diary—a year which started with great promise, with an almost 'green' summer and what seemed an ideal early autumn break in March. But since then we have experienced a very dry and foggy winter, a poor spring and now bush fires. Such vagaries in weather are common enough in Australia although here we rarely experience prolonged droughts. The bird and animal life seems to have been little affected by the dry conditions, however, and the wild flowers flourished in the spring.

The scene is quite dissimilar from this time last year. In place of the tall waving grass on the hills and the green flats all the pasture is now close-cropped and there are bare patches of ground on the rising country. Even on the better watered High Plains many of the sphagnum moss beds seem bone dry and are white and crunchy underfoot.

From a personal point of view the writing of this diary has been an exciting and revealing experience. Those things I have so often observed, consciously and sometimes subconsciously, have had to be photographed and recorded—an exercise in self-discipline and concentration that has not always been easy. But it has emphasised to me the multitude of fascinating changes and natural processes which normally tend to remain unseen or ignored.

BIBLIOGRAPHY

Adamson, C. L. 'Economic geology—Toolong diggings'. *Annual Report*, NSW Department of Mines, 1950, p. 101.
Andrews, A. *First settlement of the Upper Murray*. Ford, Sydney, 1930.
Andrews, E. C. 'Report on the Kiandra Lead'. *Geological Survey of NSW*, 1901.
Armstrong, J. A. 'Biotic pollination mechanisms in the Australian flora—a review'. *New Zealand Journal of Botany*, vol. 17, 1979, pp. 467–508.
Ashton, D. H. 'Fire in the open forests (wet sclerophyll forests)'. *Fire and the Australian Biota*, Australian Academy of Science, 1981.
Atsatt, P. R. 'Mistletoe leaf shape: a host morphogen hypothesis'. *The Biology of Mistletoes*, Academic Press, Sydney, 1983.
Australian Academy of Science. 'A report on the condition of the mountain catchment of New South Wales and Victoria'. 1957.
Baker, G. L. 'Wingless grasshoppers'. AGFACTS, 2nd edn. NSW Department of Agriculture, 1981.
Beattie, A. J. 'Ants and gene dispersal in flowering plants'.
Berenhardt, P. 'Floral biology of *Amyema*'. *The Biology of Mistletoes*, Academic Press, Sydney, 1983.
Blackbourne, B. 'Plant galls'. *Victorian Naturalist*, vol. LIV, Aug. 1937.
Borch, C. 'Life histories of some Victorian Lycaenids'. *Victorian Naturalist*, vol. XLV, Nov. 1928.
Bowen, G. D. 'Coping with low nutrients'. *The Biology of Australian Plants*, University of Western Australia Press, 1981.
Brooker, M. G. & Ridpath, M. G. 'The diet of the wedge-tailed eagle, *Aquila audax*, in Western Australia'. *Australian Wild Life Research*, vol. 7, 1980, pp. 433–52.
Bryant, W. G. 'The effect of grazing and burning on a mountain grassland, Snowy Mountains, NSW'. *Journal of Soil Conservation Service of New South Wales*, 1973.
Burr, Malcolm. *The Insect Legion*. James Nisbett, London, 1954.
Byles, B. U. 'A reconnaissance of the mountainous part of the River Murray catchment in New South Wales'. Commonwealth Forestry Bureau, bulletin no. 13, 1932.
Cadwallader, P. L. 'The role of trout as sport fish in impounded waters of Victoria and New South Wales'. Paper presented at symposium on 'Dammed Waters' held during the 1st Annual Conference of the Australian Society of Limcology, Tallangatta, 19–20 May 1979.
——. 'Some causes of the decline in range and abundance of native fish in the Murray–Darling River system'. *Proceedings of the Royal Society of Victoria*, vol. 90, 1978, pp. 211–24.
Calder, D. M. 'Mistletoes in focus: an introduction'. *The Biology of Mistletoes*, Academic Press, Sydney, 1983.
——. 'Quick on the trigger'. *Parkwatch*, no. 133, Winter 1983.
——. 'Plants do it their way'. *Parkwatch*, no. 127, Summer 1981.
Carmody, Jean. *Early days of the Upper Murray*. Shoestring Press, Wangaratta, Vic., 1981.

Carne, J. E. 'Reported damage to the Upper Murray River flats by mining operations'. *Annual Report*, NSW Department of Mines, 1902.
Carne, P. B. 'The characteristics and behaviour of the sawfly, *Perga affinis affinis* (Hymenoptera)'. *Australian Journal of Zoology*, vol. 10, no. 1, March 1962, pp. 1–34.
Carr, S. G. M. 'The role of shrubs in some plant communities of the Bogong High Plains'. *Proceedings of the Royal Society of Victoria*, vol. 75, 1962.
Carr, S. G. M. & Turner, J. S. 'The ecology of the Bogong High Plains I. The environmental factors and the grassland communities'. *Australian Journal of Botany*, vol. 7, 1959.
——. 'The ecology of the Bogong High Plains II. Fencing experiments in grassland'. *Australian Journal of Botany*, vol. 7, 1959.
Christensen, P., Recher, H. & Hoare, J. 'Responses of open forests (dry sclerophyll forests) to fire regimes'. *Fire and the Australian Biota*, Australian Academy of Science, 1981.
Cochrane, G. R., Fuhrer, B. A., Rotherham, E. R. & Willis, J. H. *Flowers and Plants of Victoria*. A. H. & A. W. Reed, Sydney, 1968.
Coleman, Edith. 'Pollination of *Pterostylis acuminata* R.Br. and *Pterostylis falcata* Rogers'. *Victorian Naturalist*, vol. L, 1934.
Common, I. F. B. 'A study of the ecology of the adult Bogong Moth (*Agrotis infusa*) (Boisd) (Lepidoptera: Noctuidae) with special reference to its behaviour during migration and aestivation'. *Australian Journal of Zoology*, vol. 2, 21 Sept. 1954, pp. 223–63.
——. 'Migration and gregarious aestivation in the Bogong Moth *Agrotis infusa*'. *Nature*, vol. 170, 6 Dec. 1952, p. 981.
Common, I. F. B. & Waterhouse, D. F. *The Butterflies of Australia*. 2nd edn. Angus & Robertson, Sydney, 1981.
Corbett, L. & Newsome, A. 'Dingo society and its maintenance: a preliminary analysis'. CSIRO, Division of Wildlife Research, Canberra.
Costermans, Leon. *Native Trees and Shrubs of South-Eastern Australia*. Rev. edn. Rigby, 1983.
Costin, A. B. 'Management opportunities in Australian high mountain catchments'. International symposium on forest hydrology, 1967.
——. 'Vegetation of high mountains in Australia in relation to land use'. *Biogeography and Ecology in Australia*, vol. 8, Sept. 1959.
Costin, A. B., Gay, L. W., Wimbush, D. J., & Kerr, D. 'Studies in catchment hydrology in the Australian Alps III'. CSIRO, Division of Plant Industries, Technical Paper no. 16, 1961.
——. 'Studies in catchment hydrology in the Australian Alps IV. Interception by trees of rain, cloud and fog'. CSIRO, Division of Plant Industries, Technical Paper no. 16, 1961.
Costin, A. B., Gray, M., Totterdell, C. J. & Wimbush, D. J. *Kosciusko Alpine Flora*. CSIRO, Melbourne, and William Collins, Sydney, 1979.
Costin, A. B. & Wimbush, D. J. 'Preliminary snow investigations'. CSIRO, Division of Plant Industries, Technical

Paper no. 15, 1961.
Costin, A. B., Wimbush, D. J. & Kerr, D. 'Studies in catchment hydrology in the Australian Alps II. Surface run-off and soil loss'. CSIRO, Division of Plant Industries, Technical Paper no. 14, 1960.
CSIRO. *The Insects of Australia*, Melbourne University Press, 1970.
Costin, A. B., Wimbush, D. J., Kerr, D. & Gay, L. W. 'Studies in catchment hydrology in the Australian Alps I. Trends in soils and vegetation'. CSIRO, Division of Plant Industries, Technical Paper no. 13, 1959.
Daley, Charles. 'The Bogong Moth'. *Victorian Naturalist*, vol. XLVIII, March 1932.
Daubenmire, R. 'Alpine timberlines in the Americas and their interpretation'. Butler University, USA, Botany Studies II, 1954?, pp. 119–36.
Dunning, D. C. 'Warning sounds of moths'. *Z. Tierpsychol.* vol. 25, pp. 129–38.
Dunning, D. C. & Roeder, K. D. 'Moth sounds and the insect-catching behaviour of bats'. *Science, NY*, 1965, pp. 147, 173–74.
ECOS 'Burning question in the Snowy'. No. 11, 1977.
_____. 'More trouble for gum trees (infestation of eucalypts by *Armillaria*)'. No. 15, Feb. 1978.
_____. 'On the trail of the rabbit'. No. 18, November 1978.
ECOS/CSIRO. 'Requiem for a rural gum tree'. No. 19, Feb. 1979.
Edwards, I. J. 'The ecological impact of pedestrian traffic on alpine vegetation in Kosciusko National Park'. *Australian Forestry*, vol. 40, 1977, pp. 108–20.
Eisner, Thomas. 'Chemical defense against predation in arthropods'. Siondheimer, E. & Simcone, J. B. (eds), *Chemical Ecology*, Academic Press, New York, 1970, pp. 157–217.
Ellis (née Sharland), Beverly Ann. 'Diet selection of two native and two introduced herbivores in an Australian rangeland region.' School of Zoology, University of New South Wales, 1975.
Ewart, A. J. *Flora of Victoria*. Macmillan, Melbourne, and Victorian Government Printer for Melbourne University Press, 1930.
Farb, Peter. *The Insects*. Life Nature Library, Time-Life International, (Nederland) NV, 1964.
Filson, R. B. & Rogers, R. W. *Lichens of South Australia*. Government Printer, South Australia, 1979.
Findlay, G. P. & Findlay, Nele. 'Anatomy and movement of the column in *Stylidium*'. *Australian Journal of Plant Physiology*, vol. 2, 1975, pp. 597–621.
Fleay, David. 'The brown snake—dangerous fellow'. *Victorian Naturalist*, vol. LIX, Jan. 1943.
Flood, J. M. The moth hunters: investigation towards a prehistory of the south-eastern highlands of Australia. PhD thesis, Australian National University, Canberra, 1973.
Frith, H. J. (ed.). *Birds in the Australian High Country*. Rev. edn. Angus & Robertson, Sydney, 1984.
Fuhrer, Bruce. *Australian Fungi*. Five Mile Press, Hawthorn, Melbourne, 1985.
Galbraith, Jean. *Wildflowers of South-East Australia*. William Collins, Glasgow, 1977.
Gill, A. M. 'Coping with fire'. *The Biology of Australian Plants*. University of Western Australia Press, 1981.
Gill, A. M., Groves, R. H., Leigh, J. H., Price, P. C., & Wimbush, D. J. 'Fire in Kosciusko National Park'. CSIRO, Division of Plant Industries, 1975.
Good, R. B. 'A preliminary assessment of erosion following wildfires in Kosciusko National Park, New South Wales, in 1973'. *Journal of the Soil Conservation Service of New South Wales*, 1973.
Grehan, J. R. 'Morphological changes in the three-phase development of *Aenetus virescens* larvae (Lepidoptera: Hepalidae)'. *New Zealand Journal of Zoology*, vol. 8, 1981, pp. 505–14.
Griffiths, Mervyn, Barker, R. & McLean, L. 'Further observations on the plants eaten by kangaroos and sheep grazed together in the paddock in south-western Queensland'. *Australian Wild Life Research*, vol. 1, 1974, pp. 27–43.
Grose, R. J. 'Some silvical characteristics and notes on the silviculture of alpine ash (*Eucalyptus delegatensis*) R. Baker'. Forestry Commission of Victoria, 1961.
Hancock, W. A. *Discovering Monaro: A Study of Man's Impact on his Environment*. Cambridge University Press, Cambridge, England, 1972.
Helms, R. 'Report on the grazing leases of the Mount Kosciusko Plateau'. *Agriculture Gazette*, NSW, vol. 4, 1893.
Hill, Gerald F. 'Tasmanian grass grub (*Oncopera intricata*, Walker)'. Council for Scientific and Industrial Research, pamphlet no. 11, Melbourne, 1929.
Hinton, H. E. 'Myrmecophilous Lycaenidae and other Lepidoptera—a summary'. *Proceedings and Transactions*, South London Entomological and Natural History Society, 1949–50, pp. 111–75.
Hughes, R. D. *Living Insects*. William Collins, Sydney, 1975.
Ingold, C. T. *The Biology of Fungi*. Hutchinson, London, 1961.
Journet, A. R. P. 'Insect herbivory on the Australian woodland eucalypt, *Eucalyptus blakelyi* M'. *Australian Journal of Ecology*, vol. 6, 1981, pp. 135–38.
Keith Turnbull Research Institute. *1983–84 Annual Report*, 1984.
Key, K. H. L. & Day, M. F. 'The physiological mechanism of colour change in the grasshopper. (*Kosciuscola tristis*) Sjost. (Orthoptera: Acrididae)'. *Australian Journal of Zoology*, Nov. 1954.
_____. 'A temperature-controlled physiological colour response in the grasshopper, *Kosciuscola tristis* Sjost. (Orthoptera: Acrididae)'. *Australian Journal of Zoology*, Nov. 1954.
Knutson, M. Donald. 'Physiology of mistletoe parasitism and disease responses in the host'. *The Biology of Mistletoes*, Academic Press, Sydney, 1983.
Land Conservation Council, Victoria. *Report on the North-Eastern Area (District 1)*. 1972.
Liddy, John. 'Dispersal of Australian mistletoes: the Cowiebank study'. *The Biology of Mistletoes*, Academic Press, Sydney, 1983.
McArthur, A. G. 'Fire behaviour in eucalypt forests'. Leaflet no. 107, Ninth Commonwealth Forestry Conference, India, 1968.
McCausland, I. P. 'Liver fluke in dairy cattle'. *Agnote*, Department of Agriculture, Victoria, June 1982.
Main, Barbara York. *Spiders*. 2nd edn. William Collins, Sydney, 1984.
Malicky, H. 'New aspects of the association between lycaenid larvae (Lycaenidae) and ants (Formicidae, Hymenoptera)'. *Journal of the Lepidopterists Society*, vol. 24, no. 3, 1970, pp. 190–202.
Marks, G. C., Fuhrer, B. A., & Walters, N. E. M. *Tree Diseases in Victoria*. Ed. Marion L. Huebner. Forests Commission Victoria, 1982.
Mascord, Ramon. *Australian Spiders in Colour*. A. H. & A. W. Reed, French's Forest, NSW, 1983.
Matthews, E. G. & Kitching, R. L. *Insect Ecology*. 2nd edn. University of Queensland Press, 1984.
Mitchell, T. W. *Corryong and the 'Man from Snowy River' district*. Wilkinson Printers for R. Boyes, Albury, 1981.
Morland, R. T. 'Erosion survey of the Hume catchment area, Parts I–V'. *Journal of the Soil Conservation Service of New South Wales*, 1958–60.
Moye, D. C. *Historic Kiandra*. Cooma–Monaro Historical Society, 1959.
New, T. R. *A Biology of Acacias*. Oxford University Press,

Melbourne, 1984.
Newsome, A. E., Corbett, L. K., Best, L. W. & Green, B. 'The dingo'. *AMRC Review*, 14-1-11, October 1973.
Nicholls, W. H. *Orchids of Australia*. Thomas Nelson (Australia), 1969.
Nolan, I. F. 'The European rabbit flea in Victoria'. Keith Turnbull Research Institute, pamphlet no. 67, 1977.
Parsons, W. F. *Noxious Weeds of Victoria*. Inkata Press, Melbourne, 1973.
Paton, J. M. *Tooma—The Centenary of Education: 137 years of settlement 1839-1976*.
Ramsbottom, John. *Mushrooms and Toadstools*. William Collins, London, 1953.
Rayment, Tarlton. 'Notes on the pollination of trigger plants'. *Victorian Naturalist*, 1948.
Rodd, N. W. 'Some observations on the biology of Stephanidae and Megalyridae (Hymenoptera)'. *Australian Journal of Zoology*, vol. 11, 2, 1951, pp. 341-46.
Rowe, R. K. *A Study of the Land in the Victorian Catchment of Lake Hume*. Soil Conservation Authority, 1967.
Rowley, Ian. *Bird Life*. William Collins, Sydney, 1975.
Rural Research CSIRO. 'Bursting the bloat bubble'. No. 78, Dec. 1972.
_____. 'Exploiting insect perfumes'. No. 101, Dec. 1978.
_____. 'Genetic blow fly control. State of the science'. No. 122, Autumn 1984.
_____. 'Phalaris for drought resistance'. No. 125, Summer 1984-85.
_____. 'The Snowy: conservation and water'. No. 11, 1973.
_____. 'Worms, flies and sheep'. No. 89, Dec. 1975.
Shields, Oakley. 'Hilltopping. An ecological study of the summit congregation behaviour of butterflies on a southern California hill'. *Journal of Research on the Lepidoptera*, vol. 6, no. 2, 1967, pp. 69-78.
Simpson, K. N. G. 'Feeding of the yellow-tailed black cockatoo on cossid moth larvae inhabiting *acacia* species'. *Victorian Naturalist*, vol. 89, Feb. 1972.
Slack, Adrian. *Carnivorous Plants*. Reed, 1980.
Slater, Peter. *A Field Guide to Australian Birds*. Vols 1-2, Rigby, 1972-74.
Staff, Ian A. & Waterhouse, John T. 'The biology of arborescent monocotyledons, with special reference to Australian species'. *The Biology of Australian Plants*, University of Western Australia Press, 1981.
Strahan, Ronald. *Complete Book of Australian Mammals*. Angus & Robertson, Sydney, 1983.
Taylor, A. C. 'Snow lease management'. *Journal of the Soil Conservation Service of New South Wales*, 1956.
Thorn, L. B. 'Notes on the life histories of some Victorian Lycaenid butterflies'. *Victorian Naturalist*, vol. XLI, July 1924.
Tillyard, R. J. *The Insects of Australia and New Zealand*. Angus & Robertson, Sydney, 1924.
Tindale, N. B. 'Note on the body temperature of a hepialid moth. (*Trictena*)'. *Records of South Australian Museum*, vol. 5, 1935, pp. 331-32.
Van Rees, H. & Beard, J. A. 'Seasonal variation in *in vitro* digestibility and chemical composition of a range of alpine plants, from Victoria, Australia'. *Australian Rangeland Journal*, vol. 6, no. 2, 1984, pp. 86-91.
Wakefield, W. A. 'Bushfire frequency and vegetational change in south-eastern Australian forests'. *Victorian Naturalist*, vol. 87, June 1970.
Willis, J. H. *Baron von Mueller and Other Pioneer Botanists of the North-East*. Murdock House, Wangaratta, 1980.
_____. *Victorian Toadstools and Mushrooms*. Field Naturalist's Club of Victoria, Melbourne, 1957.
Wilson, Edward O. 'The insect societies. The natural history of the primitive ants of the genus *Myrmercia*'. Belknap Press of Harvard University Press, Cambridge, Massachusetts, USA.
Young, Tony. *Common Australian Fungi*. New South Wales University Press, Kensington, NSW, 1982.

Index

Aborigines 19, 21, 25, 32, 36, 37, 45, 46, 53, 89
Acacia buxifolia see Box-leaf Wattle
Acacia dealbata see Silver Wattle
Acacia kettlewelliae see Buffalo Wattle
Acacia melanoxylon see Blackwood
Acacia obliguinervia see Mountain Hickory Wattle
Acacia pycnantha see Golden Wattle
Acacia rubida see Red-stem Wattle
Acacia siculiformis see Dagger Wattle
Acacia ulicifolia see Juniper Wattle
Acacia verniciflua see Varnish Wattle
Acrida conica see Long-headed Grasshopper
Acripeza reticulata see Mountain Grasshopper
Aenetus ligniviren see Apple Hanging Moth
Aenetus paradiseus montanus 122
Africa 42, 70, 137
Agarastidae 132
agarics 41, 53
Agrotis infusa see Bogong Moth
Albury 9, 75, 89
alga (*Treatepohlia aurantiaca*) 75
algae 50, 75
Alpine Ash (*Eucalyptus delegatensis*) 13, 24, 34, 50, 51, 115, 141
Alpine Beard-heath (*Leucopogon maccraei*) 99
alpine cicadas 21
alpine grasshopper (*Kosciuscola tristis*) 49
alpine grasshopper (*Monistria sp.*) 49
alpine grasshoppers 49
Alpine Marsh Marigold (*Caltha introloba*) 16
Alpine Orites (*Orites lancifolia*) 128
Alpine Oxylobium (*Oxylobium alpestre*) 13, 121, 128
Alps 21, 37, 98
Amanita 53
Amegilla sp. 35
Amphipterygidae 118
Amyema linophyllum 66
Amyema pendula see Drooping Mistletoe
Anguillaria dioica see Early Nancy
Anigozanthos sp. see kangaroo paw
Anoplognathus sp. 113, 140
ant (*Crematogaster sp.*) 110
ant (*Iridomyrmex sp.*) 99
ant lion (*Periclystus circuiter*) 120
Anthelid moth caterpillar (*Chenuala heliaspis*) 139
Anthelidae 35, 100
Antheraea eucalypti see Emperor Gum Moth
anthocyanins 22
Anthrenus sp. 58
ants 32, 36, 77, 89, 109, 120, 126, 135
Apiomorpha conica 105
Apis mellifera see introduced bee
Apple Box (*Eucalyptus bridgesiana*) 12, 43, 52, 61, 131, 135
Apple Hanging Moth (*Aenetus ligniviren*) 114
Arclotheca calendula see Capeweed
Arctiidae see tiger moths

Armidale, NSW 113
Armillaria luteobubalina 53
Ascomycetes (Discomycetes) 47
Aseroe rubra 35–36
Asilidae see robber flies
Asura lydia 100
ash 57
Australian Admiral butterfly (*Vanessa itea*) caterpillar 52
Australian Alps see Alps

bacterial fleece rot 9
Bago Plateau 74
Balcombe, William 74
Banksia marginata see Silver Banksia
banksias 43, 80
barley grass 43, 101, 126
bats 46, 100
beard-heath (*Leucopogon sp.*) 89
beard-heath (*Leucopogon biflorus*) 89, 96
Bedfordia arborescens see Blanket-leaf
bees 16, 20, 89, 99, 122, 135, 141; see also introduced bees
Beef Steak Fungus (*Fistulina hepatica*) 47
bee-flies (Bombyliidae) 99
beetle (*Anoplognathus sp.*) 113, 140
beetle (*Chrysolina sp.*) 9
beetle (*Diphucephala sp.*) 114, 124
beetle (*Metriorrhynchus sp.*) 39
beetles 4, 10, 14, 16, 49, 54, 60, 107, 112, 120, 124
Bertya findlayi 137
Big Dargal 63
Billy Buttons (*Craspedia glauca* sensu lat) 13, 16, 17, 121
Black Cicada (*Psaltoda moerens*) 7
black cockatoos see Yellow-tailed Black Cockatoo
Black Cypress-pine (*Callitris endlicheri*) 46, 80
Black Jack 3
blackberry (*Rubus fruticosus agg.*) 7, 10, 48, 61, 111
Blackberry Leaf Rust (*Phragmidium violaceum*) 48
Blackberry Orange Rust (*Kuehneola uredinis*) 48
Black-fronted Dotterel 26
Black-tongue Caladenia orchid (*Caladenia congesta*) 97
Blackwood (*Acacia melanoxylon*) 12, 81, 137
Blakely's Red Gum (*Eucalyptus blakelyi*) 12 see also red gum
Blanket-leaf (*Bedfordia arborescens*) 12, 42
bloat 81–82
blue gums (Eurabbie) 53
Blue Heeler 37
bluebells 45, 109
Blue-winged Shoveler 26, 134
Blunt Greenhood orchid (*Pterostylis curta*) 93
Bogong Moth (*Agrotis infusa*) 36–37; feasts 31, 36–37
bolete fungi 40, 41
Boletus edulis 41
Boletus satanas 41
Bombyliidae see bee-flies
Bonaparte, Napoleon 74
borers 4, 5
boronia (*Boronia nana*) 49
bossiaea 64, 114, 122, 128
Bossiaea foliosa see Leafy Bossiaea
Bothriochloa macra see red grass
Box-leaf Wattle (*Acacia buxifolia*) 81, 88

Brachychiton populneus see Kurrajong
Brachycome decipiens 121
Brachycome scapigera 122
Brachycome spathulata 121, 122
Brachyseme 43
bracken 40, 64
bracket fungus (*Crepidotus sp.*) 48
braconid wasps 58
Braconidae 58
Bradneys Gap 53, 121
briars see Sweet Briar
Bright, Vic. 9
Bringenbrong 3, 12, 32, 70
Bringenbrong State Forest 130, 133, 135, 138
Bringenbrong station 32
Broad-leaved Peppermint (*Eucalyptus dives*) 12, 93, 103
Brolga 68
brome grass (*Bromus sp.*) 43, 117, 118
Brown Hawk 34, 80, 91
Brown-headed Honeyeater 43
Buffalo Wattle (*Acacia kettlewelliae*) 102
bull ant (*Myrmecia sp.*) 15, 35, 89
Buloke (*Casuarina luehmanni*) 66
burnet moths 113
Burr, Malcolm 137
Bursaria lasiophylla 39, 136
bush fly (*Musca vetustissima*) 119–120
bush pea (*Pultenaea sp.*) 64, 98, 102
buttercups (Ranunculaceae) 121, 122, 128
butterflies 7, 16, 58, 96–97, 107, 112, 128, 135; larvae 4, 58, 86

Cabramurra 70, 72, 74, 75, 76, 82, 139, 140, 141
Caladenia carnea see Pink Fingers Orchid
Caladenia congesta see Black-tongue Caladenia orchid
Caladenia dilatata 102
caladenias 96
Callitris endlicheri see Black Cypress-pine
Callitris columellaris see White Cypress-pine
Calochilus robertsonii see Common Beard-orchid
Caltha introloba see Alpine Marsh Marigold
Calytrix tetragona see Common Fringe-myrtle
Camphor Laurel 45
Canberra 5, 79
Candalides hyacinthinus see Common Dusky Blue butterfly
Candle Heath (*Richea continentis*) 21, 128
Candlebark (*Eucalyptus rubida*) 3, 12, 22, 42, 64
Capeweed (*Arclotheca calendula*) 36, 63
Carduus tenuiflorus see Slender Thistle
Carruthers Peak 98
Carthamus lanatus 54
carpet moth (*Euphyia sp.*) 16
case moths (Pyschidae) 81; caterpillars 107
Casuarina luehmanni see Buloke
caterpillar (Agarastidae: *Comocrus behri*) 132
caterpillars 7, 16, 20, 35, 57, 58, 70, 124
caterpillars (Lasiocampidae) 132
cats 69, 88, 99, 118

cattle 2, 9, 18, 22–23, 24, 26, 34, 39, 40, 41, 42, 43, 45, 47, 52, 54, 55, 60, 64, 81, 82, 89, 102, 115, 116, 118, 123, 126, 128, 134, 137; mustering 23, 64, 128; runs 22; water for 7, 26; yards 60
cattlemen 102
cecaria see liverfluke
Cep (*Boletus edulis*) 41
Cerambycidae 5
Chauliognathus lugubris see soldier beetle
Cheleptrix felderi 47
Chenuala heliaspis see Anthelid moth caterpillar
Cherry Ballart see Wild Cherry
chestnuts 53
Chile 51
Chiloglottis gunnii see Common Bird-orchid
China Walls 58
Chlorociboria aeruginascens 47
Christmas beetle (*Anoplognathus sp.*) 113, 140
Christmas beetles 79, 101; larvae 101
Chrysididae 136
Chrysolina sp. 9
Chrysolopus spectabilis 34
chrysomelid beetle (*Poropsis augusta*) 140
chrysomelid beetles 4, 113, 139
Chrysomelidae 4, 113, 139
Chrysothrix candelaris 75
cicada (*Diemeniana euronotiana*) 129
Cladonia chlorophaea 50, 51
Clavariaceae see coral fungi
Clavulina sp. 47 see also coral fungi
clematis (*Clematis aristata*) 122
clover 23, 43, 46, 60, 80, 81, 82, 92, 94, 134
Clover Flat 19, 102, 121, 122
coccids see scale insects
Cochrans Gap 55, 98, 117
cockatoos 37, 54, 58, 78, 80, 122
cockchafer beetle (*Diphucephala sp.*) 114, 124
cockchafer larvae (Scarabaeoidea) 68
cockroaches 64
Common Beard-heath (*Leucopogon virgatus*) 99
Common Beard-orchid (*Calochilus robertsonii*) 109
Common Bird-orchid (*Chiloglottis gunnii*) 102
Common Brown butterfly (*Heteronympha merope merope*) 124, 125
Common Dusky Blue butterfly (*Candalides hyacinthinus*) 96
Common Fringe-lily (*Thysanotus tuberosus*) 109
Common Fringe-myrtle (*Calytrix tetragona*) 43, 107, 111
Common Grass Blue butterfly (*Zizina labradus labradus*) 34
Common Ground Fern (*Culcita dubia*) 46
Comocrus behri 132
Cooaninnie 63–64
Cooma 11, 24
copper sulphate 41
Copperhead 35, 121
coral fungi (Clavariaceae) 47
coral heath (*Epacris microphylla*) 49, 87, 98
Coral-pea (*Hardenbergia violacea*) 64, 81, 85
cordyceps 54

147

Cordyceps gunnii 54
Coreidae see plant bugs
correa (*Correa reflexa*) 43, 61
Correae Brown butterfly (*Oreixenica correae*) 16
Corryong, Vic. 8, 66, 75, 105
Corryong Creek 105, 126
Cortinarius 53
Cortinarius austrovenetus 53
Corybas hispidus 49
Craspedia glauca sensu lat see Billy Buttons
Crematogaster sp. see ant
Crepidotus sp. 48
Crimson Rosella 26, 37, 58, 82, 102
crows 2, 11, 47, 109, 134
crutching 9
CSIRO 10, 55, 113
cuckoo wasps (Chrysididae) 136
Cudgewa Creek 126
Culcita dubia see Common Ground Fern
cup fungi see Ascomycetes
cup fungus (*Chlorociboria aeruginascens*) 47
cup fungus (*Peziza sp.*) 47
cup moth caterpillar (*Doratifera sp.*) 111, 112

Dagger Wattle (*Acacia siculiformis*) 81
daisies 16, 121, 122, 128, 129, 139
daisy (*Brachycome decipiens*) 121
daisy (*B. spathulata*) 121, 122
damsel flies 120, 140
dandelions 21
daphne 88
daphnin 88
Dargal Creek 63, 64
Dargal Mountain 2, 3, 22, 34, 35, 37
Dargals, the 2, 3, 61, 63, 74, 105, 117, 120, 123, 137
Daviesia latifolia see Hop Bitter-pea
Daviesia ulicifolia see Gorse Bitter-pea
deciduous trees 22, 40, 57
Delias 93
Delias harpalyce see Imperial White butterfly
Depression 20
Derwent Speedwell (*Veronica derwentiana*) 138–139
Diaea sp. see flower spider
Diamond Weevil (Curculionidae: *Chrysolopus spectabilis*) 34
Dicksonia antarctica 46
Didymuria violescens 115
dieback 42
Diemeniana euronotiana 129
Diggers Speedwell (*Veronica perfoliata*) 13
Dingo (*Canis familiaris dingo*) 19, 20, 37
Diphucephala sp. 114, 124
Dipodium punctatum see Hyacinth Orchid
Diptera 61
Discomycetes see Ascomycetes
Diuris maculata 96
Diuris pedunculata see Golden Moths Orchid
docks 117
Dodder-laurel 96
dogwoods 39, 61, 111, 114
Doratifera sp. 111, 112
Double-banded Finch 75
dragon flies 26–29, 120
dragon flies (Amphipterygidae) 118
dragon fly (*Diphlebia sp.*) 118
drenching 9, 40, 41
Drooping Mistletoe (*Amyema pendula*) 65–66, 93
Drosera peltata see Pale Sundew
Drosera sp. see sundew
drought 4, 7, 24, 26, 45, 69, 113, 125, 126, 138, 143; 1982–83 4, 7, 10, 26, 40, 52, 66, 92, 119, 125, 131

Early Nancy (*Anguillaria dioica*) 93
early settlers 32, 74
Eastern Grey Kangaroo 7, 40, 46, 59, 119
Eastern Ringed Xenica butterfly (*Geitoneura acantha acantha*) 124

Eastern Rosella 54, 135
Eastern Spinebill 43
Eastern Swamphen 68
Echidna 136
Echium lycopsis see Paterson's Curse
elephant beetles see Diamond Weevil
elms 32
Embreys Lookout 105
Emperor Gum Moth (*Antheraea eucalypti*) 104–105; caterpillar 104, 105
Emus 7, 74
Epacris microphylla see coral heath
Epacris paludosa see Swamp Heath
Ephemeroptera see mayflies
erosion 10, 14, 23, 37, 43, 55, 72, 73, 93, 100, 102
Eucalyptus blakelyi see Blakely's Red Gum, red gum
Eucalyptus bridgesiana see Apple Box
Eucalyptus camaldulensis see River Red Gum
Eucalyptus camphora see Mountain Swamp Gum
Eucalyptus dalrympleana see Mountain Gum
Eucalyptus delegatensis see Alpine Ash
Eucalyptus dives see Broad-leaved Peppermint
Eucalyptus macrorhyncha see Red Stringybark
Eucalyptus pauciflora see snow gum
Eucalyptus radiata see Narrow-leaved Peppermint
Eucalyptus rossi see White Gum
Eucalyptus rubida see Candlebark
Eucalyptus viminalis see Manna Gum
Euloxia sp. 114
Eumenidae 57–58
Euphyia sp. see carpet moth
Eupoecila australasiae see Fiddler Beetle
Eurabbie 53
Europe 42, 48, 113
European Rabbit Flea (*Spilopsyllus cuniculi*) 10, 78
everlastings 13, 34, 111, 113
Exocarpus cupressiformis see Wild Cherry
Exoneura sp. 35
explorers 36

facultative diapause 36
Farrans Lookout 30, 31, 34, 94
Fasciola hepatica see liverfluke
Fergusonina sp. 61
ferns 57, 64
Fiddler Beetle (*Eupoecila australasiae*) 136
Findlay, James 137
fires 4, 24–5, 35, 45, 51, 65, 78, 103, 115, 123, 129, 130, 132, 133, 134, 135, 138, 139, 141, 142, 143
fish 32, 99
Fistulina hepatica see Beef Steak Fungus
Flame Robin 68
flies 9, 15, 36, 39, 60, 86, 119–120, 129
flies (Tachinidae) 120
floods 14, 32, 85, 86, 88
flower chafer (*Polystigma punctata*) 136
flower spider (*Diaea sp.*) 128, 129
flower wasp (*Hemithynnus hyalinatus*) 101, 109
flower wasps (Thynninae) 101
fly (*Fergusonina sp.*) 61
Fly Agaric 53
Fomes 19
foot and mouth disease 40
Forest Bronzewing 109
Forest Phebalium (*Phebalium squamulosum alpinum*) 114, 122, 128
foxes 61
Freeling, Lt-Col 74
fringe-lily (*Thysanotus sp.*) 109
fringe-myrtle see Common Fringe-myrtle
frogs 120
frost heaving 38, 48

fungi 18, 19, 35, 36, 41, 42, 50, 53, 54, 60, 114
fungus (*Phytopthora cinnamomi*) 42
fungus (*Pleurotus nidiformis*) 41, 42

Gahnia sp. see sword grass
Galah 14, 26, 32, 129, 131, 135
Galerina hypnorum 54
gall-forming insects 60–61, 105
galls 60–61, 105
Gang-gang Cockatoo 58, 94–95
Gastrimargus musicus 137
Geitoneura acantha acantha 124
Gentianella diemensis see Mountain Gentian
Geometridae 129, 132
Glossodia major see Wax-lip Orchid
glycine (*Glycine sp.*) 45
gnats 54, 103
goannas 109
gold, gold-rushes 9, 64, 74–75, 76, 77
Golden Everlasting (*Helichrysum bracteatum*) 107, 113, 124
Golden Moths Orchid (*Diuris pedunculata*) 98, 128
Golden Wattle (*Acacia pycnantha*) 86
Goodenia hederacea alpestris 122
Gorse Bitter-pea (*Daviesia ulicifolia*) 121
Goulburn 74
Grantite Buttercup (*Ranunculus graniticola*) 122
Graphium Macleayanum Macleayanum 16
grass flower (*Stipa sp.*) 3
grass seed 8, 32
Grass Trigger-plant (*Stylidium graminifolium*) 13, 128, 140–141
grasshoppers 7, 14, 48–49, 137, 138
grass-tree (*Xanthorrhoea australis*) 43, 45–46, 65, 115
Green Tree-hopper (*Sextius virescens*) 4, 15
Green-comb Spider-orchid (*Caladenia dilatata*) 102
greenhood orchid (*Pterostylis sp.*) 93
greenhood orchid (*Pterostylis x ingens*) 93
Greg Greg Fire Tail 42, 49, 54, 57
Greg Greg run 9
grevillea (*Grevillea jephcottii*) 105
Grey Currawong 102
Grey Fantail 25, 26
Grey Mare Range 63, 117
grubs 14
guinea-flower (*Hibbertia sp.*) 49, 64
Gundagai 74
Guthega catchment 72

Hakea microcarpa see Small-fruited Hakea
Handsome Flat-pea (*Platylobium formosum*) 64, 98, 99
hanging moth (*Aenetus paradiseus montanus*) 122
Hardenbergia violacea see Coral-pea
hares 118
Harrison, George 63
Harrison, Jack 63
hawk moth (*Hippotion celerio*) 29; caterpillar 28, 29
hawk moth (*H. scrofa*) 87
hawks 131
heath 43, 89, 102
heifers see cattle
Helichrysum acuminatum see Orange Everlasting
Helichrysum bracteatum see Golden Everlasting
Helichrysum leocopsideum see Satin Everlasting
Helichrysum rutidolepis see Pale Everlasting
Helipterum anthemoides 13
helmet orchid (*Corybas hispidus*) 49
Hemithynnus hyalinatus see flower wasp
Hepialidae see swift moths
herbicides 10, 46, 93
Hesperilla donnysa 86

Hesthesis sp. 39
Heteronympha merope merope see Common Brown butterfly
Hibbertia sp. see guinea-flower
High Plains 12, 23, 24, 25, 143
hill-topping 96–97
honeybee see introduced bee
honeyeaters 43, 46, 126
Hop Bitter-pea (*Daviesia latifolia*) 13, 115
hops 115
hovea 98
Hovea lanceolata see Lance-leaf Hovea
hover-flies 87, 99
humming birds 142
huntsmen spiders 68, 116
huntsmen spiders (Sparassidae) 116
Hyacinth Orchid (*Dipodium punctatum*) 124–125
hydro-electricity 21, 23
Hypericum perforatum 9
Hypochrysops byzos hecalius 124
Hypochrysops delicia delos 109, 110

ibis 70
Ichneumonid wasp 100
Ichneumonidae 100
Ictinus Blue butterfly (*Jalmenus ictinus*) 15; pupa 15
Illawarra Flame Tree 45
Imperial White butterfly (*Delias harpalyce*) 92, 93; caterpillar 92, 93; pupa 92, 93
India 93
insecticides 9
introduced bee (*Apis mellifera*) 16, 20, 107, 112
Iridomyrmex purpureus see meat ant
Iridomyrmex sp. see ant
irrigation 23, 70
Isoptera see white ants
Isotoma axillaris 45
Ivy Goodenia (*Goodenia hederacea var. alpestris*) 122

Jalmenus ictinus 15
jasmine 29, 88
Jephcott, Edwin 105
Jephcott, Sydney 105
jewel beetle (*Stigmodera sp.*) 39
jewel beetles 134
Jingellic 86, 130, 133, 135, 141
jumper ant (*Myrmecia nigrocincta*) 35, 36, 123
Juniper Wattle (*Acacia ulicifolia*) 81
Junonia villida calybe 7

Kangaroo Grass (*Themeda australis*) 2, 12, 39, 43, 59, 64, 128, 142
kangaroo paw (*Anigozanthos sp.*) 43
kangaroos 7, 8, 18, 32, 43, 46, 47, 64, 69, 78, 102, 119, 137; communication systems 54–55
Kenmore 74
Kestrel 102, 115, 131
Khancoban 3, 39, 41, 47, 63, 138
Khancoban station 30, 32
Kiandra 11, 39, 47, 53, 74, 75, 76, 98, 138
Kiandra Gold Mining Company 76
Kiandra gold-field 74–75
King Parrot 26
Kings Cross 74, 76
kookaburras 58, 102, 121, 126
Kosciuscola tristis see alpine grasshopper
Kosciusko National Park 7, 16, 19, 20, 23, 37, 39, 40, 70, 78, 86, 100, 101, 103, 111, 119, 133
Kuehneola uredinis see Blackberry Orange Rust
Kunzea ericifolia 139
Kunzea parvifolia see Violet Kunzea
kunzeas 61, 88, 107, 111, 128
Kurrajong (*Brachychiton populneus*) 36, 45

Labium sp. 20
lacewings (Neuroptera) 115, 120
lambs see sheep
Lance-leaf Hovea (*Hovea lanceolata*) 81
Larentiinae 103
Lasiocampidae 132

Lasioglossum sp. 122
Late Black Wattle 5
Latrodectus mactans 142
leafhoppers *see* Green Treehopper
Leafy Bossiaea (*Bossiaea foliosa*) 114, 128
Leioproctus sp. 20
Leopard Orchid (*Diuris maculata*) 96
Leptorhynchos tenuifolius see Scaly Buttons
Leptospermum brevipes see Slender Tea-tree
Leptospermum glabrescens see Smooth Tea-tree
Leptospermum grandifolium see Mountain Tea-tree
Leptospermum multicaule see Silver Tea-tree
lerp insects (Psyllidae) 79
Leucopogon biflorus see beard heath
Leucopogon maccraei see Alpine Beard-heath
Leucopogon sp. see beard-heath
Leucopogon virgatus see Common Beard-heath
lichen (*Chrysothrix candelaris*) 75
lichen (*Cladonia chlorophaea*) 50, 51
lichen (*Parmelia rutidota*) 50, 51
lichen (*Protoparmelia petraeoides*) 50, 51
lichens 22, 50–51, 75
Ligar (surveyor) 75
Ligars Route 75
Lighthouse crossing 31–32
Lighthouse homestead 31
Lighthouse Mountain 2, 3, 8, 10, 43, 45, 52, 57, 66, 112, 115, 119, 124, 125, 135, 136, 137
lightning strikes 25, 103, 129, 130, 135
Liquidambar 57
liverfluke (*Fasciola hepatica*) 40–41
lizards 14, 115, 121
Lobs Hole gold-field 74
Long-headed Grasshopper (*Acrida conica*) 5
longicorn beetle (Cerambycidae) larvae 5
longicorn beetle (*Hesthesis sp.*) 39
longicorn beetle larvae 46
long-tailed wasps (Megalyridae) 5
looper caterpillars (Geometridae) 129, 132
lorikeets 43
Lucilia cuprina 8–9
Lycaenid larvae 15
Lycaenidae 15
Lycosidae 68–69
Lymantriidae 112
lyrebirds 54, 58, 64, 113, 118

McInnes family 77
Macleay's Swallowtail (*Graphium macleayanum macleayanum*) 16
Macquarie Perch 32
magnolia family 122
Magpie-lark 80
magpies 26, 70, 94, 123, 126, 135
malaria 53
'Man From Snowy River, The' 3
Manjar 3
Manna Gum (*Eucalyptus viminalis*) 42
Mantidae 115, 133
mantispas (Mantispidae) 115
Marasmius sp. 47
march flies (*Scaptia maculiventris*) 18
Maroonhood orchid (*Pterostylis pedunculata*) 93
mason wasps (Eumenidae) 57–58; larvae 58
mayflies (Ephemeroptera) 54, 99, 115
Meadow Argus butterfly (*Junonia villida calybe*) 7
meat ant (*Iridomyrmex purpureus*) 4, 15, 107, 134, 136, 137–138
medic burr 36, 134
Mediterranean 55
Megalyridae 5
Melbourne 15, 52, 77, 110, 116, 118, 138
Melichrus urceolatus see Urn Heath
Metriorrhynchus sp. 39
mice 115

Microseris lanceolata 21
Middle East 70
midges 49
mimicry 39, 54, 58, 87, 113, 115
mistletoe (*Amyema linophyllum*) 66
Mistletoe-bird 65, 66
mistletoes 65–66, 93, 132
mites 139
Mittamatite 6, 105, 124, 125
Mocatta 46
Mocattas Ridge 64, 65, 100
molybdenum 80
Monistria sp. see alpine grasshopper
monotremes 136
Moonlight Jewel butterfly (*Hypochrysops delicia delos*) 109, 110; caterpillar 109, 110
mosquitoes 11, 49, 54, 78, 93, 103, 121
moss beds *see* sphagnum moss
mosses 15, 54, 57
moth caterpillars 4, 57, 79, 100, 120, 122, 139
moth (*Cheleptrix felderi*) 47
moth (*Euloxia sp.*) 114
moth (*Pollanisus sp.*) 113
moths 10, 16, 35, 36, 39, 46, 54, 58, 60, 70, 100, 103, 104, 105, 107, 114, 133, 142
moths (Larentiinae) 103
moths (Lymantriidae) 112
Mount Bogong, Vic. 37
Mount Burrow 69, 89, 105
Mount Elliott 62, 63, 68, 77
Mount Elliott gold-field 77
Mount Jagumba 3, 137
Mount Jagungal 3, 35, 74, 75, 114, 128
Mount Kosciusko 2, 8, 31, 35, 74, 77, 98, 105
Mount Mitta Mitta 77
Mount Townshend 98
Mount Twynam 98
Mountain Gentian (*Gentianella diemensis*) 20, 21, 35
Mountain Grasshopper (*Acripeza reticulata*) 39, 40
Mountain Gum (*Eucalyptus dalrympleana*) 12, 13, 22, 39, 123
Mountain Hickory Wattle (*Acacia obliguinervia*) 34, 102
Mountain Pepper (*Tasmannia lanceolata*) 122, 129
Mountain Swamp Gum (*Eucalyptus camphora*) 12, 53, 120, 123, 137
Mountain Tea-tree (*Leptospermum grandifolium*) 122
mud dauber wasps (Sphecidae) 136
mules operation 9
Murray Cod 32
Murray pine *see* White Cypress-pine
Murray Plateau 3, 75
Murray River 2, 3, 12, 30, 31, 32, 50, 57, 63, 75, 82, 83, 86, 88, 91, 94, 105, 110, 112, 117, 124, 125, 126, 128, 143
Murray River flats 3, 12, 32, 70, 85, 106
Murray River valley 2, 8, 30, 35, 70
Musca vetustissima see bush fly
museum beetle (*Anthrenus sp.*) 58
mushrooms 18, 41
musk 42
Mycena interrupta 53
mycorrhizae 53, 124–125
Myrmecia nigrocincta see jumper ant
Myrmecia sp. see bull ant
Myrtaceae 20
myxomatosis 10–11, 78, 91, 117, 118

Narrow-leaved Peppermint (*Eucalyptus radiata*) 12, 40, 42, 46, 47, 53, 59, 64, 78
National Park *see* Kosciusko National Park
National Parks Service 23
native bee (*Amegilla sp.*) 35
native bee (*Exoneura sp.*) 35
native bee (*Lasioglossum sp.*) 122
native bee (*Leioproctus sp.*) 20
native bees 107, 112
native bluebells 45, 109
native violets 49
native Yam (*Microseris lanceolata*) 21

nematode parasites 36, 61
nettles 52
Neuroptera *see* lacewings
New Caledonia 89
New Chum Hill mine 76
'New Chum' reef 77
New Zealand 114
Nodding Blue-lily (*Stypandra glauca*) 105
Nodding Greenhood orchid (*Pterostylis nutans*) 93
North America 42
Nymphalidae 7

oaks 57
Ogilvies Creek 13, 16, 102
Orange Everlasting (*Helichrysum acuminatum*) 13, 16
orchids 95, 102, 121
Oreixenica correae 16
Orichora Brown butterfly (*Oreixenica orichora orichora*) 129
Orites lancifolia see Alpine Orites
oxylobium 122
Oxylobium alpestre see Alpine Oxylobium

Pale Everlasting (*Helichrysum rutidolepis*) 13, 16
Pale Sundew (*Drosera peltata*) 83–85, 103
Papua New Guinea 70
paraquat 93
parasites, parasitism 4, 5, 15, 20, 42, 44, 50, 52, 53, 54, 57, 58, 65, 66, 86, 96, 120, 125, 136
Parmelia rutidota 50, 51
parrots 43, 116
Paterson, A. B. (Banjo) 3
Paterson, Mrs 9
Paterson's Curse (*Echium lycopsis*) 9–10
peat 21, 22
peppermints 3, 141
Perennial Rye Grass 55
Perga affinis 16, 18
Perga sp. 79
Periclystus circuiter see ant lion
Peziza sp. 47
Phalaenoides glycine see vine moth caterpillar
phalaris 43, 55
Phasmatid (*Didymuria violescens*) 115
phebalium *see* Forest Phebalium
pheromones 15, 105, 118, 138
Phigaloides Skipper butterfly (*Trapezites phigaloides*) 97
photosensitivity *see* wort dermatitis
Phragmidium violaceum see Blackberry Leaf Rust
Phytopthora cinnamomi 42
Pied Cormorant 82
Pied Currawong 70, 82, 102
Pierces Fire Trail 59
Pierces Hut 128
Pimelea linifolia 88
Pimelea sp. see rice-flower
Pine Mountain 89, 105
pines 53, 77
Pink Fingers Orchid (*Caladenia carnea*) 82, 96
Pink-bells (*Tetratheca ciliata*) 96, 99, 121
Pixies Parasol (*Mycena interrupta*) 53
plant bugs (Coreidae) 10, 107, 139
Platylobium formosum see Handsome Flat-pea
Platypus 136
Playle, Arnold 126
Pleurotus 41–42
Pleurotus nidiformis 41, 42
Poa sp. see snowgrass
Pollanisus sp. 113
Polystigma punctata 136
Polyura pyrrhus sempronius 48
pomaderris 12, 42, 45, 46, 57, 61, 124
poplars 32, 57, 72
Poropsis augusta 140
Possum Point 57
possums 66
predators 9, 10, 14, 20, 39, 44, 48, 52, 63, 66, 69, 99, 100, 112, 118, 120, 122, 133, 137
preying mantis (Mantidae) 115, 133

Protoparmelia petraeoides 50, 51
Psaltoda moerens see Black Cicada
Pseudoperga sp. 25
Pseudozethus sp. 39
Psyllidae 79
Pterostylis acuminata 93
Pterostylis curta see Blunt Greenhood orchid
Pterostylis x ingens 93
Pterostylis nutans 93
Pterostylis pedunculata see Maroonhood orchid
Pterostylis sp. see greenhood orchid
Pultenaea sp. see bush pea
Purple Violet (*Viola betonicifolia*) 121
Pyschidae *see* case moths

Queensland 46

rabbits 9, 10, 11, 39, 40, 61, 64, 69, 78, 91, 115, 117, 118, 119, 126, 135
Rainbow-bird 14, 100
Ramaria sp. 47 *see also* coral fungi
Ranunculaceae *see* buttercups
Ranunculus graniticola 122
red grass 43, 66
red grass (*Bothriochloa macra*) 43, 62
red gum (*Eucalyptus blakelyi*) 12, 38, 45, 53, 79, 116, 130
red gums 2, 38, 39, 43, 44, 45, 78, 79, 85, 113, 117, 130, 131, 134, 137
Red Kangaroo 87
Red Stringybark (*Eucalyptus macrorhyncha*) 3, 12, 42, 43, 66, 81, 88, 93, 103
Red-back Spider (*Latrodectus mactans*) 142
Red-bellied Black Snake 120
Red-browed Finch 26, 94
Red-necked Wallaby 59, 87
Red-rumped Parrot 14, 26, 94, 95
Red-stem Wattle (*Acacia rubida*) 45, 86
rice-flower (*Pimelea sp.*) 36, 88, 89, 109
Richea continentis see Candle Heath
River Red Gum (*Eucalyptus camaldulensis*) 3, 12, 32, 53, 106
robber flies (Asilidae) 112, 128
Rock Isotome (*Isotoma axillaris*) 45
Rosa rubiginosa see Sweet Briar
Round Mountain 48, 122, 123, 128, 129
rove beetles (Staphylinidae) 86
Rubus fruticosis agg. see blackberry
Rubus parvifolius see Small-leaf Bramble
Ruby Bracket Fungus (*Tyromyces pulcherrimus*) 18–19
Russula 53
rust 48

Sacred Ibis *see* White Ibis
Saffron Thistle (*Carthamus lanatus*) 54
St John 9
St John's Wort (*Hypericum perforatum*) 9
St Helena 74
Salvation Jane *see* Paterson's Curse
saprophytes 42
sarcoptic mange 40
Sarsaparilla *see* Coral-pea
Satin Everlasting (*Helichrysum leocopsideum*) 16
savannah woodland 12
sawfly (*Perga affinis*) 18; larva 16, 18
sawfly (*Perga sp.*) 79; larva 79
sawfly (*Pseudoperga sp.*) 25; larva 25
scale insects 4, 15, 60, 61, 89
Scaly Buttons (*Leptorhynchos tenuifolius*) 16
Scaptia maculiventris 18
scarab beetles (Scarabaeoidea) 101
sclerophyll forest, dry 12, 25, 103; wet 12
scribble gum 70, 72
sedges 13

149

Senecio lautus see Variable Groundsel
Sextius virescens see Green Tree-hopper
Sharp Greenhood orchid (*Pterostylis acuminata*) 93
shearing 101, 134
sheep 8, 9, 11, 19, 24, 25, 26, 34, 39, 40, 41, 43, 45, 47, 55, 61, 63, 64, 70, 95, 100, 101, 112, 115, 116, 126, 134, 137; water for 7, 26
sheep blowfly (*Lucilia cuprina*) 8-9
Sherbrooke Forest, Vic. 118
Short-horned Grasshopper (*Acrida conica*) see Long-headed Grasshopper
Silver Banksia (*Banksia marginata*) 42, 43, 45, 64
silver birches 53
Silver Tea-tree (*Leptospermum multicaule*) 96
Silver Wattle (*Acacia dealbata*) 4-5, 12, 15, 34, 64, 77, 81, 119
Silybum marianum 112
skink lizards 133
skipper butterfly (*Hesperilla donnysa*) 16
skippers (Hesperiidae) 16, 96
Slender Rice-flower (*Pimelea linifolia*) 88
Slender Tea-tree (*Leptospermum brevipes*) 111-112
Slender Thistle (*Carduus tenuiflorus*) 112, 129
Small-fruited Hakea (*Hakea microcarpa*) 98
Small-leaf Bramble (*Rubus parvifolius*) 111-112, 137
Smooth Tea-tree (*Leptospermum glabrescens*) 120
snakes 14, 120-121
Snakey Plains 22, 35, 37, 121, 123
Snakey Plains Track 22, 24, 34, 37, 121
snow gum (*Eucalyptus pauciflora*) 18, 22, 24, 37, 48, 70-72, 76, 128, 129, 140, 141
snow gums 3, 13, 16, 18, 21, 22, 24, 25, 34, 35, 37, 39, 72, 74, 76, 102, 114, 122, 123, 128, 139, 140, 141
snow leases 22, 23, 64
snow-fields 21, 74, 82
snowgrass (*Poa sp.*) 13, 16, 20, 23, 35, 37, 39, 40, 64, 113, 122, 139
Snowy Mountains 3, 24, 72, 134
Snowy Mountains Authority 11, 24, 39, 72
Snowy Mountains Scheme 23, 72
sodium fluoroacetate (1080) 10, 11, 47
sod-seeding 93
Soil Conservation Authority 23
soldier beetle (*Chauliognathus lugubris*) 21
South Africa 9
South America 42
South Australia 9, 46
South-East Asia 70
sphagnum moss (*Sphagnum cristatum*) 20-21, 35, 128, 143
Sphecidae 136
sphecoids 20
Spice Islands, Indonesia 42
spider orchids 82
spiders 15, 20, 58, 66, 112, 116, 136
Spilopsyllus cuniculi see European Rabbit Flea

Spilosoma curvata 101
Spring Tops 59
Starfish Fungus (*Aseroe rubra*) 35-36
starlings 109, 116
Stigmodera sp. 39
Stipa sp. see grass flower
stock see cattle, sheep
stock routes 42, 64, 122
stockmen 23
Straw-necked Ibis 70
Striated Thornbill 37
stringybarks 81, 141, 142
Strzelecki, *Count* 98
Stuckey's station 74
Stylidium graminifolium see Grass Trigger-plant
Stypandra glauca 105
sub clover 43, 80
Sue City 82
Sulphur-crested Cockatoo 2, 14, 25-26, 58, 59, 118, 119, 135
sundew (*Drosera sp.*) 83-85, 101
Superb Blue Wren 26, 68, 138
superphosphate 40, 43, 80
swamp gum see Mountain Swamp Gum
Swamp Heath (*Epacris pauludosa*) 13
Swamp Wallaby (*Wallabia bicolor*) 40
swans 68
Sweet Briar (*Rosa rubiginosa*) 7, 94
swift moth (*Trictena argentata*) 44
swift moths (Hepialidae) 20, 44, 114; caterpillars 44, 54
sword grass (*Gahnia sp.*) 86
Sydney 77

Tachinidae 120
Tailed Emperor butterfly (*Polyura pyrrhus sempronius*) 48
Talbingo Dam 74
Talbingo Hill 75
Tasmania 21, 46, 128
Tasmannia lanceolata see Mountain Pepper
Taylors Creek 15, 46, 58, 64, 109, 134
Taylors Creek valley 46, 70, 110
tea-trees 20, 39, 61, 96, 102, 109, 111, 137
termites 15
terrestrial orchids 82, 102
Tetratheca ciliata see Pink-bells
The Insect Legion 137
Themeda australis see Kangaroo Grass
thistles 2, 26, 32, 46, 58, 131
Thowgla Creek 126
Three Mile Dam 76
thrips 60
Thymelaeaceae 88
Thynninae 101
Thysanotus patersonii see Twining Fringe-lily
Thysanotus tuberosus see Common Fringe-lily
tiger moth (*Asura lydia*) 100
tiger moth (*Spilosoma curvata*) 101
tiger moth (*Utetheisa pulchelloides*) 100
tiger moths (Arctiidae) 100, 111, 112
Tintaldra 91
toadstool (*Armillaria luteobubalina*) 53
toadstool (*Cortinarius austrovenetus*) 53

toadstool (*Marasmius sp.* formerly *Collybia elegans*) 47
toadstools 18, 41, 42, 53, 54
Tom Groggin station 2
Toolong Gold Field 23, 63-64
Tooma 86
Tooma Dam 13, 102, 114, 115, 122
Tooma River 2, 3, 13, 30, 46, 56, 57, 59, 63, 65, 70, 73, 88, 95, 117, 124, 125, 128, 135
Tooma River flats 70, 74, 85, 91, 106
Tooma River valley 6, 8, 30, 69
Towong 63, 68, 69, 84, 86
Towong Bridge 31, 68, 82, 83, 94, 110, 111, 117, 143
Towong Hill station 30, 32, 128, 137
transmission line 11, 39, 63, 86, 87, 98
Trapezites phigaloides 97
Treatepholia aurantiaca see alga
treecreepers 135
treefern (*Dicksonia antarctica*) 46
Trictena argentata 44
trigger-plant see Grass Trigger-plant
Trout Cod 32
Tufted Daisy (*Brachycome scapigera*) 122
Tumbarumba 9, 74
Tumut 75
Tumut River 74, 75
Tumut 2 Power Station 72
tussock grass 20, 23, 64, 74
Twiggy Mullein (*Verbascum virgatum*) 117
Twining Fringe-lily (*Thysanotus patersonii*) 109
Tyromyces pulcherrimus 18-19

Upper Murray 2, 4, 74, 82, 86, 130
Urn Heath (*Melichrus urceolatus*) 49
Utetheisa pulchelloides 100

Vanessa itea see Australian Admiral butterfly
Variable Groundsel (*Senecio lautus*) 105, 129
Variegated Thistle (*Silybum marianum*) 112
Varnish Wattle (*Acacia verniciflua*) 81, 102
vegetable caterpillar (*Cordyceps gunnii*) 54
Verbascum virgatum 117
Veronica derwentiana 138-139
Veronica perfoliata 13
Victoria 8, 9, 10, 48, 75, 81
Victorian High Plains 23
vine moth caterpillar (*Phalaenoides glycine*) 113, 123
Viola betonicifolia 121
Violet Kunzea (*Kunzea parvifolia*) 43, 88, 103, 107
von Mueller, *Baron* Ferdinand 10, 105, 137

Wallabia 40
Wallabia bicolor see Swamp Wallaby
wallabies 16, 59, 87, 137
wallaby grass 43
wasp (*Labium sp.*) 20
wasp (*Pseudozethus sp.*) 39

wasps 15, 39, 54, 60, 86, 87, 100, 135
wattles 15, 39, 53, 57, 61, 66, 80, 81, 82, 95, 102, 109, 114
Wax-lip Orchid (*Glossodia major*) 96
Wedge-tailed Eagle 10-11, 47, 112
weeds 7, 9, 10, 32, 46, 48, 63, 64, 88, 101, 112, 117, 142
weevils (Curculionidae) 4
Welcome Swallow 54, 113, 142
Welumba Creek 3, 43, 54, 57, 61, 80, 82, 89, 94, 95, 96, 102, 107, 109, 110, 111, 113, 118, 121, 124
Welumba Creek waterfall 57, 61, 88
Welumba Hill 58
Western Australia 42, 47
Western Grey Kangaroo 46-47
Wheelers Hut 23
Whistling Eagle 135
white ants (Isoptera) 46
White Chamomile Sunray (*Helipterum anthemoides*) 13
White Cypress-pine (*Callitris columellaris*) 46
White Egret 68, 85
White Gum (*Eucalyptus rossii*) 70
White Ibis 70, 110
Whitehead family 31
White-necked Heron 106
Wild Cherry (*Exocarpus cupressiformis*) 52-53
wild dogs 8, 19-20, 37, 40, 70, 100, 111, 118, 134
wild pigs 40
Wild Raspberry see Small-leaf Bramble
Willie Ploma 74
willows 7, 32, 50, 57, 70, 72, 73-74
Willy Wagtail 14, 47-48, 112, 113, 118, 123, 131
wolf spiders (Lycosidae) 68-69
Wolseleys Gap 34
wombats 39, 40, 64
Wood Duck 26, 106, 110
Wood Swallow 142
wood-boring caterpillars 4-5
Woolly bear caterpillars (Anthelidae) 35, 100
Woollybutt see Alpine Ash
World's End 3
wort dermatitis 9

Xanthorrhoea australis see grass-tree

yacca gum 46
Yarrangobilly River 74
Yellow Boy 137
Yellow Boy Creek 137
Yellow Robin 49
Yellow Spot Jewel butterfly (*Hypochrysops byzos hecalius*) 124
Yellow-tailed Black Cockatoo 4, 37, 78, 80, 122
Yellow-tailed Thornbill 88
Yellow-tufted Honeyeater 54
Yellow-winged Locust (*Gastrimargus musicus*) 137
Youngal 3

Zizina labradus labradus 34